# SIDREA Series in Accounting and Business Administration

This is the official book series of SIDREA - the Italian Society of Accounting and Business Administration. This book series is provided with a wide Scientific Committee composed of Academics by SIDREA. It publishes contributions (monographs, edited volumes and proceedings) as a result of the double blind review process by the SIDREA's thematic research groups, operating at the national and international levels. Particularly, the series aims to disseminate specialized findings on several topics – classical and cutting-edge alike – that are currently being discussed by the accounting and business administration communities. The series authors are respected researchers and professors in the fields of business valuation; governance and internal control; financial accounting; public accounting; management control; gender; turnaround predictive models; non-financial disclosure; intellectual capital, smart technologies, and digitalization; and university governance and performance measurement. This book series is indexed in Scopus.

Antonio De Vito • Francesco Grossetti

# Tax Avoidance Research

## Exploring Networks and Dynamics of Global Academic Collaboration

Springer

Antonio De Vito
University of Bologna
Bologna, Italy

Francesco Grossetti
Bocconi University
Milan, Italy

ISSN 2662-9879 ISSN 2662-9887 (electronic)
SIDREA Series in Accounting and Business Administration
ISBN 978-3-031-51764-8 ISBN 978-3-031-51765-5 (eBook)
https://doi.org/10.1007/978-3-031-51765-5



This Springer imprint is published by the registered company Springer Nature Switzerland AG
The registered company address is: Gewerbestrasse 11, 6330 Cham, Switzerland

*To Madrid, where it all started.*

# Acknowledgments

We sincerely want to express our gratitude to the colleagues who supported and coped with our commitment to writing this monograph. In particular, we would like to thank our past and current coauthors: Olga Bogachek, Elisa Casi, Carlo D'Augusta, Paul Demere, Martin Jacob, and Claudia Imperatore. Your understanding and collaboration were invaluable during this journey. We also thank Salvatore Mastrullo for excellent research assistance. This work would not have been possible without the resources and support of our home institutions, the University of Bologna and Bocconi University.

# Contents

# List of Figures

# List of Tables

# Chapter 1
# Introduction

In accounting, understanding the intricate dynamics of tax avoidance has garnered significant attention from scholars worldwide (Hanlon & Heitzman, 2010; Dyreng & Maydew, 2017; Wilde & Wilson, 2018).[1] This monograph is a comprehensive and systematic overview of global tax avoidance research conducted over the past two decades.[2] Its primary objective is to provide academics with a strong interest in tax avoidance research with valuable guidance into the prevailing trends, key contributors, and influential academic journals in this field.

Tax avoidance, a complex phenomenon at the intersection of accounting, finance, economics, and law, necessitates a multifaceted approach to exploring its various dimensions comprehensively. To this end, the research methodology employed in this monograph leverages social network analysis (Borgatti et al., 2009), which enables the detection and interpretation of intricate and often concealed patterns within tax avoidance research networks. By analyzing the collaborative efforts of scholars, their affiliations, and the countries involved and identifying the most impactful academic journals, this study aims to shed light on the extensive web of connections that underlie tax avoidance research.

In recent years, there has been a notable upsurge in the interest surrounding tax avoidance, primarily driven by several high-profile tax scandals that have

[1] In this monograph, we adopt a comprehensive definition of tax avoidance, which encompasses the reduction of taxes through various means. This definition aligns with Dyreng et al. (2008) and encompasses all transactions that impact a firm's explicit tax liability. Throughout this work, we might employ the terms "tax avoidance," "tax aggressiveness," "tax sheltering," and "tax planning" interchangeably.

[2] To ensure a rigorous and comprehensive analysis, the decision to focus on an international sample in our literature review stems from methodological considerations aimed at addressing potential limitations and enhancing the robustness of our findings. This approach strengthens the external validity of our research and enables us to observe patterns and trends that may transcend national boundaries.



A. De Vito, F. Grossetti, *Tax Avoidance Research*, SIDREA Series in Accounting and Business Administration, https://doi.org/10.1007/978-3-031-51765-5_1

brought firms' tax practices into the limelight.[3] These scandals have captured the attention of policymakers, tax authorities, the media, and activist groups alike. Consequently, policymakers across the globe have undertaken concerted efforts (e.g., the OECD's Base Erosion and Profit Shifting (BEPS) initiative (OECD, 2018)) and have proposed new policy tools, such as the global minimum tax (OECD, 2023). Simultaneously, tax authorities have intensified their scrutiny of corporate group structures that aim to reduce tax liabilities by transferring profits from one jurisdiction to another (Klassen & Laplante, 2012; Markle, 2016). Similarly, the media and activist groups have played a pivotal role in raising public awareness regarding corporate tax avoidance practices (Dyreng et al., 2016). This growing attention to tax avoidance has engendered a climate where its implications are thoroughly examined and addressed by key stakeholders. The significant involvement of policymakers, tax authorities, the media, and activist groups underscores the critical importance of understanding and mitigating tax avoidance behavior in contemporary society.

The issue of corporate tax avoidance has generated significant interest, leading to a substantial body of academic literature. Starting from Shackelford and Shevlin (2001) and Weisbach (2001), initial studies focused on defining and measuring tax avoidance. Hanlon (2003) and McGill and Outslay (2004) attempted to estimate tax avoidance based on financial statement data, while other researchers explored tax shelters (Wilson, 2009; Lisowsky, 2010; Brown, 2011) or developed measures based on aggressive book-tax differences (Desai & Dharmapala, 2006, 2009; Frank et al., 2009). A significant contribution in this field is the work of Dyreng et al. (2008), who developed a cash-based effective tax rate measure capable of capturing various forms of tax avoidance. Interestingly, their study revealed substantial unexplained variation in long-run cash effective tax rates, prompting further investigation into firm-specific factors that may explain such cross-sectional differences in tax avoidance outcomes. A firm-centric perspective, linking corporate characteristics to tax outcomes, has emerged as a subsequent research avenue, as discussed in the comprehensive review by Wilde and Wilson (2018). The proliferation of academic research in response to the widespread interest and concern surrounding corporate tax avoidance underscores the significance of understanding the underlying dynamics and factors influencing tax outcomes in different contexts.

The monograph is organized as follows. Chapter 2 presents a comprehensive and detailed analysis of the theoretical foundations of tax avoidance, aiming to elucidate its unique characteristics in contrast to tax evasion and the complexities associated with its definition. It goes beyond a surface-level examination by delving into the various methodologies employed for measuring tax avoidance. Moreover, the chapter thoroughly investigates the multifaceted drivers contributing to tax avoidance's prevalence. In addition to exploring the determinants and measurement of tax avoidance, this chapter extensively examines the wide-ranging economic

[3] Two prominent examples of such scandals have been promptly reported by the Financial Times (i.e., Houlder (2014, 2016)). For the interested reader, we also refer to O'Donovan et al. (2019).

implications stemming from such practices. It emphasizes the impact of tax avoidance on crucial accounting aspects such as corporate transparency, the cost of capital, and firm value.

Chapter 3 offers an introduction to graph theory, providing a solid foundation for understanding network analysis's fundamental concepts and applications. It begins by probing into the definition of a graph and its key components, such as nodes and edges, which form the basis of network structures. Moreover, the section discusses the different types of graphs, including their fundamental mathematical properties such as connectivity, paths, and cycles. In addition to exploring the theoretical aspects, this section also addresses practical considerations when building networks based on real relational systems. It discusses the essential steps in building a network, emphasizing the significance of effective data gathering and representation methodologies. The section also highlights the importance of optimizing network layouts to enhance visual representation, including techniques for aligning nodes and edges within a network. Furthermore, the role of visualization in comprehending network structure and patterns is underscored, emphasizing the value of compelling visualization renderings and tools.

Chapter 4 extensively examines empirical studies conducted on tax avoidance throughout the past two decades. It investigates the process of selecting a proper sample, elucidating the criteria and methodologies employed in this regard. To comprehensively understand the data, the chapter offers some descriptive statistics that provide a holistic view of the tax avoidance research landscape. Moving beyond the initial overview, the analysis treats each observation or paper obtained through the academic Scopus Application Programming Interface (or API) as a combination of various components. These components include authors, affiliations, countries, journals, abstracts, and keywords. Exploring these diverse entities makes it possible to detect and study the underlying communities and relationships within this academic domain. Through the empirical investigation, Chap. 4 sheds light on the collaborative nature of tax avoidance research. It highlights how researchers join forces, contributing to a specific research domain and collectively advancing knowledge in the field. Furthermore, this chapter is a valuable resource for aspiring scholars with a keen interest in tax avoidance research. It offers guidance, insights, and a deeper understanding of this important study area.

Chapter 5 concludes the monograph by summarizing the key findings and insights from the empirical analysis. Understanding tax avoidance research entails acknowledging individual authors' central role and affiliations' dynamic nature, where small groups of scholars within institutions contribute to the overall centrality of universities and research organizations. Analyzing journal networks reveals trends and interrelationships among tax avoidance publications, while the field's interdisciplinary nature transcends traditional disciplinary boundaries, intersecting accounting, finance, economics, and law. It is critical to emphasize the relevance of policy implications and leverage technology and information systems as they have become essential in today's fast-paced world.

## References

Borgatti, S. P., Mehra, A., Brass, D. J., & Labianca, G. (2009). Network analysis in the social sciences. *Science, 323*(5916), 892–895.

Brown, J. L. (2011). The spread of aggressive corporate tax reporting: A detailed examination of the corporate-owned life insurance shelter. *The Accounting Review, 86*(1), 23–57.

Desai, M. A., & Dharmapala, D. (2006). Corporate tax avoidance and high-powered incentives. *Journal of Financial Economics, 79*(1), 145–179.

Desai, M. A., & Dharmapala, D. (2009). Corporate tax avoidance and firm value. *The Review of Economics and Statistics, 91*(3), 537–546.

Dyreng, S., & Maydew, E. L. (2017). Virtual issue on tax research. *Journal of Accounting Research, 56*(2). https://onlinelibrary.wiley.com/page/journal/1475679x/homepage/virtual_issue_on_tax_research_in_jar.htm

Dyreng, S. D., Hanlon, M., & Maydew, E. L. (2008). Long-run corporate tax avoidance. *The Accounting Review, 83*(1), 61–82.

Dyreng, S. D., Hoopes, J. L., & Wilde, J. H. (2016). Public pressure and corporate tax behavior. *Journal of Accounting Research, 54*(1), 147–186.

Frank, M. M., Lynch, L. J., & Rego, S. O. (2009). Tax reporting aggressiveness and its relation to aggressive financial reporting. *The Accounting Review, 84*(2), 467–496.

Hanlon, M. (2003). What can we infer about a firm's taxable income from its financial statements? *National Tax Journal, 56*(4), 831–863.

Hanlon, M., & Heitzman, S. (2010). A review of tax research. *Journal of Accounting and Economics, 50*(2), 127–178.

Houlder, V. (2014). Leak reveals scale of corporate tax deals with Luxembourg. *Financial Times.*

Houlder, V. (2016). Panama papers yield leads for UK task force, says Hammond. *Financial Times.*

Klassen, K. J., & Laplante, S. K. (2012). Are us multinational corporations becoming more aggressive income shifters? *Journal of Accounting Research, 50*(5), 1245–1285.

Lisowsky, P. (2010). Seeking shelter: Empirically modeling tax shelters using financial statement information. *The Accounting Review, 85*(5), 1693–1720.

Markle, K. (2016). A comparison of the tax-motivated income shifting of multinationals in territorial and worldwide countries. *Contemporary Accounting Research, 33*(1), 7–43.

McGill, G. A., & Outslay, E. (2004). Lost in translation: Detecting tax shelter activity in financial statements. *National Tax Journal, 57*(3):739–756.

O'Donovan, J., Wagner, H. F., & Zeume, S. (2019). The value of offshore secrets: Evidence from the Panama papers. *Review of Financial Studies, 32*(11), 4117–4155.

OECD. (2018). OECD/g20 inclusive framework on BEPS: Progress report July 2018-may 2019. Technical report, Organisation for Economic Co-operation and Development.

OECD, (2023). Tax challenges arising from the digitalisation of the economy – administrative guidance on the global antibase erosion model rules (pillar two) – OECD/g20 inclusive framework on BEPS, OECD. Technical report, Organisation for Economic Co-operation and Development.

Shackelford, D. A., & Shevlin, T. (2001). Empirical tax research in accounting. *Journal of Accounting and Economics, 31*(1–3), 321–387.

Weisbach, D. A. (2001). Ten truths about tax shelters. *Tax Law Review, 55*, 215.

Wilde, J. H., & Wilson, R. J. (2018). Perspectives on corporate tax planning: Observations from the past decade. *Journal of the American Taxation Association, 40*(2), 63–81.

Wilson, R. J. (2009). An examination of corporate tax shelter participants. *The Accounting Review, 84*(3), 969–999.

# Chapter 2
# Exploring Tax Avoidance: A Synthesis of the Literature

## 2.1 Introduction to Tax Avoidance Research

Globalization has sparked numerous structural transformations in the economic system, leading to a shift underway for several years. It involves transitioning from a localized business model focused on growth and development within the national market (*national identity*) to a global one (*global identity*) that simultaneously seeks profit across multiple markets (Desai, 2009).

In this era of globalization, corporate taxation has undergone a meticulous and sophisticated makeover with strategic implications. Consider, for example, the activities of multinational corporations operating in diverse jurisdictions. Tax management is no longer a matter of mere regulatory compliance for these companies. Instead, they adopt a proactive approach to meet regulatory obligations in specific countries while preemptively avoiding potential conflicts with tax authorities in their various markets. This shift in perspective has led to tax compliance receding in importance over time. Instead, companies have embraced ever-more intricate and assertive strategies for tax avoidance, driven solely by the desire to minimize the burden of taxes throughout the company's lifecycle, as they believe taxation to be a significant cost element that must be kept to a minimum (Donohoe et al., 2014).

Undeniably, an aggressive tax planning strategy offers substantial benefits in terms of financial advantages, such as increased cash flow to support growth and a substantial rise in net income after taxes. This, in turn, benefits shareholders by enhancing shareholder value. However, it is essential to acknowledge that such benefits come hand in hand with several costs. These costs include fees paid to external legal and tax advisors for tax planning and the potential expenses associated with disputes with tax authorities. Additionally, some costs are hard to quantify, such as the adverse direct and indirect effects on a company's reputation (Gallemore et al., 2014), the mounting political and social pressure influencing corporate decisions (Dyreng et al., 2016), and consumer boycotts of goods and services.

 
A. De Vito, F. Grossetti, *Tax Avoidance Research*, SIDREA Series in Accounting and Business Administration, https://doi.org/10.1007/978-3-031-51765-5_2

Consumers increasingly view aggressive tax planning as ethically unacceptable behavior (Hoi et al., 2013).

Considering these factors and the overarching objective of maximizing profits, companies now recognize the imperative to reassess their approach to taxation. No longer confined to the tax department, isolated from strategic decisions made by top management, taxation requires a comprehensive strategy encompassing cost and risk analysis. This is particularly crucial due to the escalating complexity, demands, and risks of tax matters. The primary objectives are twofold: minimizing tax burdens in globally conducted business operations and ensuring a high level of compliance that significantly reduces the need for tax authorities' scrutiny (Beasley et al., 2021).

The academic community has extensively explored these themes. Amidst the vast array of studies, the focus of this chapter is to examine the main contributions of accounting scholars to the understanding of tax avoidance, with particular attention to the theoretical frameworks that underpin their work. The chapter is structured into four sections, each serving a specific purpose. The initial section delves into the conceptual aspects of tax avoidance and presents a comprehensive literature review that illuminates the key theoretical elements framing the subject matter. This helps understand the relationship between tax avoidance strategies' marginal costs and benefits (Wilde & Wilson, 2018).

Section 2.3 explores the challenges associated with measuring tax avoidance, examining critical aspects such as the efficacy of applicable measurement methods and their limitations and advantages, while Sect. 2.5 describes how firms avoid taxes through income shifting and reviews the methodology developed by Grubert and Mutti (1991) and Hines Jr and Rice (1994).

Section 2.6 investigates internal and external factors that "determine" the level of tax avoidance. These factors include company size, corporate strategic decisions, management remuneration, as well as executive characteristics, among others. The emphasis will also be on factors whose comprehension is still somewhat limited but is crucial for understanding the phenomenon.

In Sect. 2.7, the research will analyze the economic consequences of corporate tax avoidance. Despite the considerable growth in research on the factors that drive tax avoidance (Hanlon & Heitzman, 2010; Wilde & Wilson, 2018), there has been a notable lack of attention given to exploring the consequences of corporate tax avoidance strategies. This is surprising for two compelling reasons. First, taxes represent a substantial expense for businesses, making it inherently attractive for firms to employ legal means to reduce their tax liabilities through tax avoidance practices (Beasley et al., 2021). However, it is important to recognize that tax avoidance comes with costs, and firms must strike a delicate balance between potential tax benefits and the associated non-tax costs (McClure, 2023). Consequently, the overall impact of tax avoidance on firms and investors may not always be positive or value enhancing (Scholes et al., 2014). Second, managers make tax avoidance decisions alongside other crucial reporting, financing, and investment choices. Although tax strategies may not always be the primary concern, their economic implications for firms' reporting, financing, and investment policies are significant enough to warrant meticulous examination by accounting scholars.

For example, taxes exert a pervasive influence on the choice between financing operations through debt or equity. Moreover, corporate investments serve as a vital driver of firm value. Assessing a project's net present value and deciding whether to invest is a function of tax payments' magnitude, timing, and unpredictability (Jacob, 2022). Importantly, these decisions regarding financing and investment have substantial implications for both firm value and investors' assessments of their anticipated after-tax returns.

Taken together, the analysis will follow the well-known "Scholes-Wolfson" (Scholes et al., 2014) theoretical framework, which emphasizes the importance of considering "*all parties, all taxes, and all costs*" in evaluating tax management decisions. Hence, the focus will be on (i) *which* firms engage in tax avoidance, (ii) *why* firms avoid taxes, i.e., the determinants of corporate tax avoidance, and (iii) *what* happens to firms that engage in tax avoidance, i.e., what economic consequences tax-avoiding firms potentially face.

## 2.2 Theoretical Background of Tax Avoidance

Before discussing the theoretical framework adopted in the present monograph, defining the tax avoidance phenomenon on a conceptual level is crucial. Tax avoidance typically refers to behaviors that occur openly without concealing taxable income. Unlike tax evasion, which constitutes a criminal offense, tax avoidance is a *behavior* that is not explicitly prohibited by the law. It involves using legal provisions and exploiting loopholes in the tax code to achieve tax savings.

Although the law distinguishes between these two phenomena, the same cannot be said from an economic standpoint. As Blouin (2014) correctly points out, an inherent empirical challenge in tax avoidance research is the difficulty in drawing the line between non-aggressive and aggressive tax avoidance activities. Hanlon and Heitzman (2010) have attempted to define and classify the phenomenon. Instead of categorizing tax strategies as either non-aggressive or aggressive, the authors suggest that tax avoidance should be viewed as a spectrum (as in a continuum) of strategies. This spectrum, depicted in Fig. 2.1, ranges from legally reducing taxes through activities such as investing liquidity in municipal bonds with favorable tax rates to engaging in abusive and high-risk tax strategies involving tax havens (e.g., Desai et al. 2006; Dyreng and Lindsey 2009). In this perspective, tax avoidance includes both certain (*legal*) and uncertain (*riskier and occasionally illegal*) tax positions that may be subject to scrutiny and challenge by the tax authority.

Because of a lack of a universally accepted definition of tax avoidance, tax scholars have thus turned to theoretical models of the economics of crime proposed by Becker (1968) and Allingham and Sandmo (1972) to describe the phenomenon. These models explore, from an individualistic and rational perspective, the incentives to evade taxes. They conclude that the extent of tax evasion is negatively correlated with the probability of detection by tax authorities, the severity of

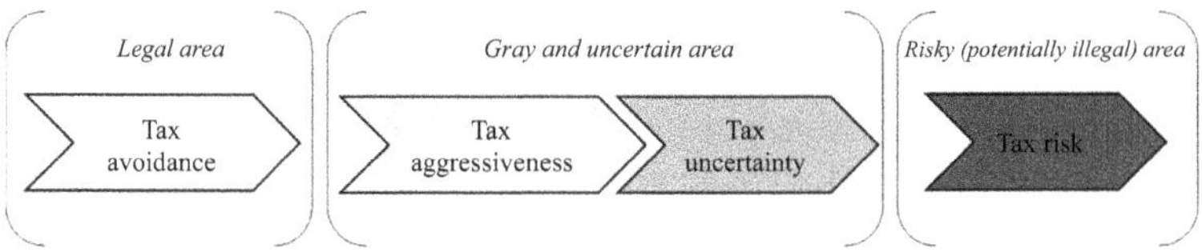


**Fig. 2.1** The tax avoidance spectrum (Alexander and De Vito 2021, p. 10)

penalties, an individual's aversion to risk, and intrinsic motivations to fulfill tax obligations.

Hence, economic theory suggests that profit-maximizing individuals should employ strategies to reduce the tax burden as long as the expected marginal benefits outweigh the associated marginal costs (Chen & Chu, 2005; Crocker & Slemrod, 2005). However, it is worth considering whether these models also apply to corporate taxpayers, given the additional complexities arising from the separation of ownership and control. Tax scholars agree that both models are suitable for examining corporate tax avoidance (Slemrod & Yitzhaki, 2002). When there is a close ownership base and there are no conflicts between the principal and the agent (commonly known as agency costs), the gains generated from tax avoidance activities, given a certain level of risk, would be distributed among all owners as if a single individual undertook the behavior. However, the situation differs for widely held companies with a clear separation between ownership and control. In such cases, conflicts of interest may arise between shareholders, who are typically risk-neutral and more inclined to accept a higher level of tax avoidance, and management, which is more risk-averse and less inclined to accept a higher level of tax avoidance due to its inherently risky nature (Slemrod, 2004). To address this, it is essential to consider the managerial incentive mechanisms necessary to minimize agency costs. As discussed in Sect. 2.3, the empirical literature has focused on various mechanisms, including the design of executive compensation contracts linked to after-tax results and, ultimately, firm value.

Overall, there is a consensus that tax avoidance enables management to minimize the tax burden and maximize profits for shareholders. However, Desai et al. (2007) also maintain that these strategies may not always benefit all shareholders equally. To effectively implement tax avoidance strategies that shield taxable income from tax authorities, management needs to employ complex and sometimes opaque corporate structures, often located in low-tax jurisdictions. These structures may divert resources for the personal benefit of management and majority shareholders at the expense of minority shareholders. This strategy, known as "income diversion" or "tunneling" in the literature, stems from divergent interests among majority and minority shareholders. The authors theoretically show that the interests of minority shareholders are aligned with those of the tax authority, as both are adversely affected by tax avoidance activities. It is also important to note that the alignment of interests is bidirectional. When the tax authority effectively reduces tax avoidance, minority shareholders would benefit indirectly as fewer resources would be diverted for the personal benefit of management and majority shareholders.

The contributions above (Chen & Chu, 2005; Crocker & Slemrod, 2005; Desai et al., 2007), albeit fundamental, primarily focus on **agency costs**, often examined in isolation. However, given the nature of the firm as a nexus of contracts between different parties (Coase, 1937), other tax avoidance costs should also be carefully considered. Wilde and Wilson (2018) propose a framework (hereafter WW) where expected marginal benefits of tax avoidance are evaluated against multiple marginal costs, consistent with a trade-off analysis. These costs include not only agency costs related to conflicts of interest between management and shareholders (or between majority and minority shareholders) but also **implementation costs** incurred in tax planning (e.g., fees paid to external legal and tax consultants for tax planning) and **outcome costs**, such as unpaid taxes and penalties if the tax authority deems a particular tax planning strategy aggressive and unlawful.

In line with the seminal works of Scholes et al. (2014) and Shackelford and Shevlin (2001), this comprehensive approach represents the conceptual framework for examining the phenomenon at hand. As shown in Fig. 2.2, the WW framework aligns with the multidisciplinary nature of tax research and with the importance of considering "*all taxes, all parties, and all taxes*" when analyzing tax avoidance strategies. In Sect. 2.3, we will refer to this conceptual framework to structure the review on the determinants of tax avoidance and to emphasize the advancements in the literature over the past two decades.

## 2.3 Measuring Tax Avoidance

This section delves into the various metrics used to measure tax avoidance in the literature. Our key argument is that not all measures suit every research question. First, we explore the sources of information for firm-level tax data, followed by a detailed examination of the most commonly utilized tax avoidance measures.

Regarding taxable income and tax payments, which are crucial in assessing tax avoidance, corporations report them on their tax returns and Generally Accepted Accounting Principles (GAAP) financial statements. Tax returns are not publicly available and limited to a few researchers, so most tax avoidance measures rely on financial statement data. However, it is widely recognized that estimating taxable income from financial statements poses numerous challenges. Hanlon (2003) and McGill and Outslay (2004) have highlighted potential issues associated with this process. Essentially, financial statements lack disclosure regarding taxable income and the actual cash taxes paid or anticipated for the current year's earnings. The Financial Accounting Standards Board (FASB) primarily focuses on establishing rules for financial accounting standards that enable external stakeholders to evaluate a company's economic performance and forecast future performance. FASB's primary objective is not to disclose tax-related data, except to the extent these items impact GAAP earnings and the GAAP balance sheet. However, external parties

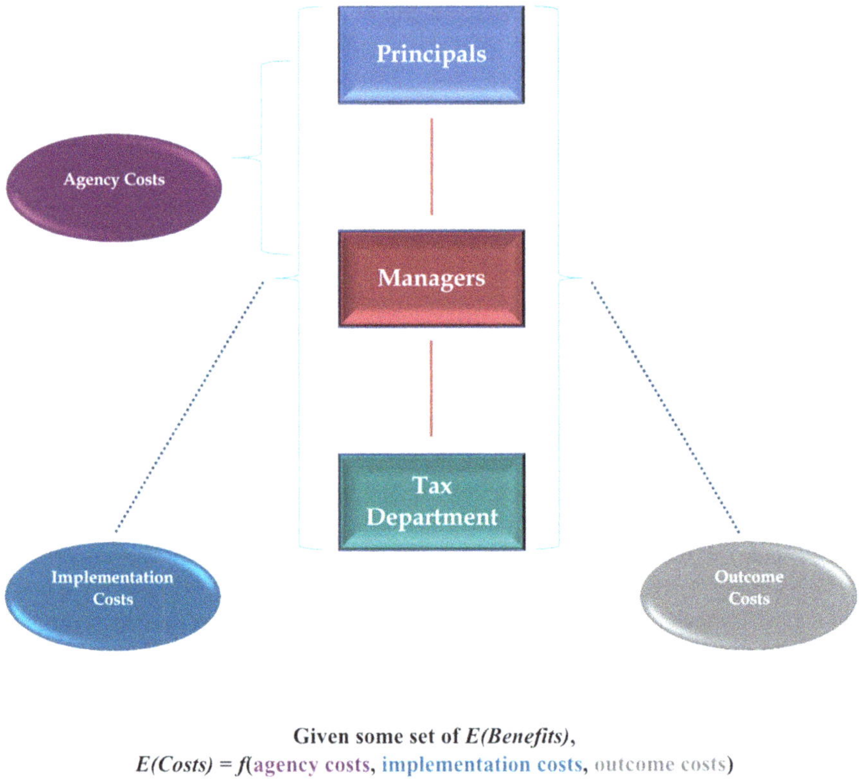


**Fig. 2.2** The framework of tax avoidance as in Wilde and Wilson (2018), p. 64

may have reasons for wanting to ascertain taxable income, such as benchmarking accounting earnings or evaluating corporate tax avoidance strategies.

Considering the limitations in estimating taxable income from financial statements, it is essential to consider what access to tax returns, as an alternative source, would and would not resolve. Due to differing book and tax reporting rules, it is nearly impossible to match any tax return to a particular set of financial statements. Even with tax return data, determining the amount of tax paid on reported accounting earnings or cash flows disclosed in a filed 10-K is challenging, if not impossible. While one could compare the tax return to the accounting earnings of entities included in the tax return, this approach may not provide a solution when benchmarking accounting earnings for a specific group of entities, such as the consolidated set for GAAP reporting. Furthermore, tax regulations and enforcement occur at the national level, so relying solely on tax returns filed in a jurisdiction would only offer information about the portion of activity in that jurisdiction for a multinational corporation. Moreover, if researchers seek the market's interpretation of a firm's taxable income, they must use data available to the market, which is something other than tax return data. Finally, research conducted using tax return data lacks replicability. While tax return data can undoubtedly prove valuable in specific contexts, this may not always be true.

### 2.3.1 *Effective Tax Rates*

Effective tax rates (ETRs) are computed by dividing an estimated tax liability by a metric representing pre-tax profits or cash flow. These metrics indicate the average tax rate per income unit or cash flow. It is crucial to understand the components of the numerator. In the GAAP ETR, the numerator is the total income tax expense. Hence, the ratio remains unaffected by tax strategies that defer taxes, such as accelerated depreciation for tax purposes. However, certain factors unrelated to tax planning, such as valuation allowance changes or tax contingency reserve adjustments, may still impact it. Moreover, the GAAP ETR directly affects accounting earnings. On the other hand, the Cash ETR is calculated using cash taxes paid in the numerator and can be influenced by tax deferral strategies but not by changes in tax accounting accruals. Notably, there may be a discrepancy between the numerator and the denominator in the yearly Cash ETR if the cash taxes paid include taxes paid on earnings from a different period (e.g., as a result of a tax authority's audit completed in the current year), but the denominator only includes current period earnings.

Depending on the research objective, the denominator of the ETR may be a more critical factor to consider. Most ETRs usually employ pre-tax GAAP earnings as the denominator, thus only capturing *non-conforming tax avoidance* (i.e., transactions treated differently for book and tax purposes). Consequently, if a firm not constrained by financial accounting regulations (e.g., a private firm) manages to avoid a significant portion of taxes by reporting lower accounting earnings and taxable income (i.e., *conforming tax avoidance*), effective tax rate measures would not capture such tax avoidance.[1] In sum, researchers need to exercise caution when drawing conclusions about tax avoidance, especially when analyzing a sample that includes firms with varying degrees of importance placed on financial accounting earnings.

Finally, Dyreng et al. (2008) have developed a long-term Cash ETR. This measure is computed by summing up the cash paid for income taxes over 10 years, scaled by the sum of pre-tax income (adjusted for special items) during the same period. The main advantage of this measure is its focus on the long term, which reduces the year-to-year volatility usually observed in yearly ETRs. By using extended periods, the measure thus minimizes the mismatch between cash taxes

[1] When book-tax conformity is high and firms do not face capital market pressure (Burgstahler et al., 2006; Bonacchi et al., 2019), we suggest using the conforming tax avoidance proxy developed by Badertscher et al. (2019).

paid and earnings.[2] The use of cash taxes paid in the numerator is advantageous as it avoids the influence of tax accrual effects found in current tax expenses.

### 2.3.2 Book-Tax Differences

Another measure of non-conforming tax avoidance developed in the literature is obtained by calculating the difference between pre-tax book income and an estimate of taxable income for a firm, where current tax expenses are placed in the numerator, and the statutory corporate tax rate is placed in the denominator (book-tax differences or BTDs). If the difference between these income measures is positive, researchers infer that the firm has engaged in tax avoidance strategies.

Mills (1998) uses various measures of book-tax differences on a sample of US firms and finds that as the difference between pre-tax book income and the estimate of taxable income increases, there is a higher likelihood of being subjected to a tax audit and audit adjustments by the US tax authority (Internal Revenue Service or IRS). Similarly, Desai (2003); Wilson (2009) use BTD measures to analyze the trend in the level of tax aggressiveness among US firms, concluding that the positive differences between book income and the estimates of taxable income during the sampled years can be attributed to aggressive tax planning.[3]

Desai and Dharmapala (2006) and Desai and Dharmapala (2009b) introduce an abnormal BTD measure by regressing total book-tax differences on total accruals to control for earnings management. The authors then use the residual of such regression as a proxy for tax avoidance behavior. Similarly, Frank et al. (2009) estimate the "discretionary" portion (DTAX) of their "PERMDIFF" measure, which represents the difference between effective and statutory tax rates multiplied by pre-tax accounting income. Similar to the Jones (1991) model of discretionary accruals, this measure attempts to isolate management's *intentional* tax avoidance actions by removing underlying business factors unrelated to tax avoidance strategies.

However, it is important to emphasize that regression-based models strictly depend on the proxies used as business factors. Since taxes influence various corporate decisions (e.g., investment and financing policies), it lies on the researcher to determine which variables should be explicitly included in the model as known business factors unrelated to tax avoidance and which effects should be left to fall into the residuals to capture *intentional* actions to avoid taxes. As Hanlon and Heitzman (2010) argue, researchers should carefully consider their research

[2] Such a mismatch could be due to (i) timing differences between actual tax payments and tax returns, (ii) settlements with the tax authority on previous fiscal periods, or (iii) taxes on repatriated earnings as is the case in the United States (Hanlon & Heitzman, 2010).

[3] Nonetheless, it is worth pointing out that Shevlin (2001) cautions against making definitive conclusions regarding the levels and patterns of tax avoidance solely based on book-tax differences.

question and the implications of their model before making inferences from the results.[4]

### 2.3.3 *Tax Shelter Activity*

Wilson (2009) uses a sample of 59 firms accused of engaging in tax shelters to investigate the impact of tax shelter participation on financial reporting. The author develops a model to analyze the characteristics of tax shelter participation and assess management's tax-planning strategies and the implications of tax shelters on shareholder wealth.

The author finds a positive association between BTDs and the occurrence of tax sheltering. This relationship holds when comparing actual tax shelter participants with size- and industry-matched control firms or a broader set of control firms. By examining court documents and financial reports, the author also computes the amount of federal tax savings generated by the tax shelters for 33 firms in the sample and finds that these tax savings contribute to an average increase of 102 percent in reported BTDs during the years of active tax shelter participation. The author also observes a positive correlation between discretionary accruals and instances of tax sheltering. However, discretionary accruals alone do not fully explain the observed BTD results, indicating that BTDs contain information about aggressive tax reporting unrelated to discretionary accruals and accrual-based earnings management. These findings demonstrate the significance of tax shelter participation in influencing reported BTDs and highlight the incremental utility of BTDs in identifying tax shelter participants.

Finally, to explore the implications of tax shelter use further, the author examines whether tax sheltering aligns with managers maximizing shareholder value. Drawing on Desai and Dharmapala (2006)'s proposition that complex tax shelter schemes may facilitate rent extraction rather than enhancing shareholder wealth, the author analyzes the stock return performance around tax shelter participation and finds that tax shelter firms with strong corporate governance scores exhibit significantly higher abnormal returns compared to tax shelter firms with poor corporate governance. These results suggest that tax sheltering can create shareholder wealth for firms with strong corporate governance practices.

---

[4] In this regard, it is worth highlighting that BTD proxies can also be used to measure book-tax conformity not only tax avoidance. For detailed reviews on this issue, we refer the reader to Menicacci (2022a,b)

### 2.3.4 Unrecognized Tax Benefits (UTBs)

Another tax avoidance measure is the UTB reserve, which is used to gauge the magnitude and fluctuations in the accounting reserve for potential tax contingencies.[5]

Tax scholars rely on UTB to examine firms that adopt risky and aggressive tax planning strategies. However, before using it to assess a firm's tax avoidance strategy, it is worth highlighting two underlying and somewhat conflicting forces that drive UTB and may make it not a clean tax avoidance measure. On the one hand, taxes undoubtedly play a prominent role, as higher UTBs imply higher uncertainty in a firm's tax positions and a heightened level of tax avoidance. On the other hand, financial reporting incentives could also affect the amount of UTBs (i.e., the amount of the accounting accrual for potential future tax audits), which is subject to management's discretion and judgment and ultimately affects the bottom-line earnings, prompting researchers to scrutinize it for signs of earnings management. Hence, researchers interested in using UTB as a proxy for tax avoidance need to be aware of the dual nature of this accounting item and include a thorough discussion of its implications for their research question, design, and results since earnings management and tax avoidance incentives could seemingly affect UTB amounts.

### 2.3.5 Effective Tax Planning

Recently, Schwab et al. (2022) have developed a measure of effective tax planning, which aligns with the SW paradigm. As explained above, the SW paradigm maintains that effective tax planning involves considering "*all taxes, all parties, and all costs*" when maximizing after-tax returns. The authors use data envelopment analysis (DEA) to construct the measure, which assesses a firm's efficiency in converting inputs into outputs. The measure incorporates inputs such as R&D expenditures, PP&E, tax haven operations, intangible assets, inventory, and leverage, enabling estimation for profit and loss firms whose tax avoidance incentives may differ.[6] The authors show their effective tax planning measure is incremental to Cash ETRs in predicting after-tax returns and captures elements of effective tax planning beyond what is captured by Cash ETRs. The authors also provide evidence that tax-effective firms reduce tax cash outflows and retain a greater portion of their uncertain tax benefits, lowering non-tax costs associated with cash flow uncertainty.

---

[5] Firms disclose UTB amounts under Financial Accounting Standards Board Interpretation No. 48, Accounting for Uncertainty in Income Taxes (FIN 48) issued in 2006.

[6] Tax avoidance studies often exclude loss firms from the analyses due to their different tax positions. Henry and Sansing (2018) have developed a measure of corporate cash tax avoidance that encompasses all observations and captures the degree to which a firm is tax-favored.

### *2.3.6 Measuring Tax Avoidance for Non-US Firms*

Most tax avoidance measures we have examined have primarily been applied to US firms. This is due to the availability of reliable information on taxes paid, which is often not easily found in the financial statements of non-US firms. Atwood et al. (2012) have developed a method to approximate the amount of taxes paid, regardless of the institutional context under investigation. The authors measure tax avoidance as the difference between the firm's "unmanaged tax amount" (the home–country statutory corporate tax rate times pre-tax income before exceptional items) and its "managed tax amount" (current taxes paid). Positive differences reflect how aggressively managers pursue strategies that reduce taxes paid.

## 2.4 A Critical Examination of Tax Avoidance Measures

The literature review has thus far identified multiple contributions proposing various measures to quantify tax avoidance. These measures, derived from financial statements, serve as useful approximations of taxable income and, ultimately, define a firm's tax avoidance.

However, considering the abovementioned measures reveals three critical issues in a firm's tax avoidance measurement process. The first issue stems from a clear definition of the phenomenon. Despite significant progress in the literature, we still lack a unified theory that systematically and distinctively defines tax avoidance. For example, while Chen et al. (2010) define tax avoidance as a strategy involving "*downward management of taxable income through tax planning activities*," Frischmann et al. (2008) (p. 265) include any "*tax positions with relatively weak supporting facts*" under tax avoidance. As we have seen in Sect. 2.2, to address this theoretical limitation, Hanlon and Heitzman (2010) propose a broader tax avoidance concept encompassing practices ranging from pure legality to near evasion.

The lack of clarity on the problem naturally raises the second issue, which concerns the choice and utilization of tax avoidance measures. As emphasized by Dunbar et al. (2010), the measures available to researchers, as presented in this work, are not strictly equivalent as they capture different dimensions of the phenomenon. As previously argued, even seemingly similar measures such as the GAAP ETR and the Cash ETR capture distinct tax avoidance strategies, with the latter considering both *temporary* and *permanent* differences between the reported book income and the estimated taxable income. Hence, in this scenario, the researcher's subjective judgment and knowledge play a significant role. To build confidence, we encourage future researchers to articulate explicitly and transparently the objectives underlying their research question, the scope of the investigation, and why they think the measures employed are appropriate to achieve the intended purpose. While subjectivity can never be eliminated, doing so would, at least, minimize it.

The third issue pertains to using these measures, initially designed for the US context, to study tax avoidance in other jurisdictions. Except for the work of Atwood et al. (2012), the lack of contributions to developing measures calibrated for different tax systems is evident. Instead, researchers use the existing measures, even though they may not be entirely suitable outside the United States. Given that tax avoidance is now a global phenomenon (OECD, 2018), we call for researchers to develop new measures specifically tailored to study tax avoidance in contexts beyond the United States, thus enhancing the reliability of findings on this phenomenon.

Before analyzing the factors that drive tax avoidance and digging into its economic consequences, based on the analysis of the literature thus far, Fig. 2.3 provides a summary and a brief description of each measure as well as the methodological approach for their calculation.

## 2.5 The Hines and Rice's (1994) Income Shifting Approach

Thus far, the analysis has primarily centered around measures assessing the realized outcomes of past tax avoidance strategies. However, before discussing the studies on the determinants and consequences of tax avoidance, the attention now shifts toward examining the mechanisms or channels through which firms engage in tax avoidance. As we have seen, firms employ various strategies to reduce their tax burdens, encompassing entirely legal and more aggressive strategies. The focus is now on a technique extensively utilized by multinational corporations to shift profits to low-tax jurisdictions, known as "income or profit shifting."

Income shifting refers to the strategic practice employed by multinational corporations to transfer profits or income generated in high-tax jurisdictions to lower tax jurisdictions, thereby reducing their overall tax liabilities. It involves implementing financial and operational arrangements within the multinational structure, such as inter-company transactions, transfer pricing, and the allocation of intellectual property rights, to shift the taxable income from higher tax to lower tax locations. The purpose of income shifting is to exploit disparities in tax rates across jurisdictions and optimize the multinational corporation's tax position by minimizing the amount of taxable income subject to higher tax rates while maximizing the allocation of income in jurisdictions with more favorable tax treatment.

The empirical estimation of profit shifting in economics primarily relies on a methodological approach derived from seminal research on multinational income shifting, particularly the influential studies by Hines Jr and Rice (1994) and Grubert and Mutti (1991).[7] These foundational studies established a conceptual framework

[7] For a survey of the empirical literature on tax-motivated income shifting within multinational firms, see, for example, Dharmapala (2014).

| Measure | Computation | Description | Impact accounting earnings? | Reflect deferral strategies? | Reflect non-conforming avoidance? | Reflect conforming avoidance? | Computable by jurisdiction? |
|---|---|---|---|---|---|---|---|
| GAAP ETR | $\frac{\text{Worldwide total income tax expense}}{\text{Worldwide total pre-tax accounting income}}$ | Total tax expense per dollar of pre-tax book income | Yes | No | Yes | No | Yes |
| Current ETR[a] | $\frac{\text{Worldwide current income tax expense}}{\text{Worldwide total pre-tax accounting income}}$ | Current tax expense per dollar of pre-tax book income | Maybe | Yes | Yes | No | Yes |
| Cash ETR[b] | $\frac{\text{Worldwide cash taxes paid}}{\text{Worldwide total pre-tax accounting income}}$ | Cash taxes paid per dollar of pre-tax book income | No | Yes | Yes | No | No |
| Long-run cash ETR[c] | $\frac{\sum(\text{Worldwide cash taxes paid})}{\sum(\text{Worldwide total pre-tax accounting income})}$ | Sum of cash taxes paid over *n* years divided by the sum of pre-tax earnings over n years | No | Yes | Yes | No | No |
| ETR Differential[d] | Statutory ETR—GAAP ETR | The difference of between the statutory ETR and the firm's GAAP ETR | Yes | No | Yes | No | No |
| DTAX[e] | Error term from the following regression: ETR differential × Pre-tax book income $= a + b \times$ Controls $+ e$ | The unexplained portion of the ETR differential | Yes | No | Yes | No | No |
| Total BTD[f] | Pre-tax book income $-$ ((U.S. $CTE + Fgn\ CTE$)/U.S. $STR$) $- (NOL_t - NOL_{t-1})$) | The total differences between book and taxable incomes | Yes for a portion, no for a portion | Yes | Yes | No | Yes (U.S.) |
| Temporary BTD | Deferred tax expense/U.S. STR | | No | Yes | Yes | No | Yes (U.S.) |
| Abnormal total BTD[g] | Residual from $BTD/TA_{it} = \beta TA_{it} + \beta m_i + e_{it}$ | A measure of unexplained total book-tax differences | Yes for a portion, no for a portion | Yes | Yes | No | No |
| Unrecognized tax benefits[h] | Disclosed amount post-FIN48 | Tax liability accrued for taxes not yet paid on uncertain positions | Yes | If uncertain | Yes, some | Yes, some | No |
| Tax shelter activity[i] | Indicator variable for firms accused of engaging in a tax shelter | Firms identified via firm disclosures, the press, or IRS confidential data | Depends on the type of shelter | Shelter may be a deferral strategy; but not an overall measure | Not overall-measure is transaction based | Not overall-measure is transaction based | Unlikely |
| Marginal tax rate[j] | Simulated marginal tax rate | Present value of taxes on an additional dollar of income | No | Yes | Yes | Yes | Not with existing data |

**Fig. 2.3** Measures of tax avoidance (Hanlon and Heitzman (2010), p. 140)

that continues to exert significant influence. According to Hines Jr and Rice (1994), the pre-tax income observed in a foreign affiliate comprises two components: the "true" income and the "shifted" income, which can be positive or negative. The true income is generated through the utilization of capital and labor inputs by the affiliate. Therefore, the analysis incorporates measures of the affiliate's capital and labor inputs, such as fixed tangible assets and employment compensation, to predict the hypothetical "true" income level.

On the other hand, the shifted income is determined by the tax incentive to transfer income between the affiliate and other entities. In its simplest form, this incentive is represented by the tax rate differential between the parent company's statutory corporate tax rate and the affiliate's statutory corporate tax rate. However, more intricate versions of the approach consider the overall tax rate structure faced by all the affiliates of the multinational corporation. For example, Huizinga and Laeven (2008) consider the broader pattern of tax rates faced by all the affiliates of the multinational corporation. The income attributed to profit shifting involves the reported income of a low-tax affiliate that cannot be adequately explained by the affiliate's labor and capital inputs.

Dharmapala and Riedel (2013) propose a novel approach to measuring profit shifting that deviates significantly from the Hines–Rice approach. Their methodology involves using exogenous changes in the parent firm's income, often referred to as "income shocks," to analyze the sensitivity of profits reported by the affiliate to such exogenous changes. Given an existing profit-shifting structure, a fraction of this exogenous income is expected to be shifted to affiliates operating in jurisdictions with lower tax rates than the parent. However, this would not apply to affiliates facing higher tax rates than the parent. Consequently, affiliates with high tax rates serve as a control group in the analysis. One challenge of this approach is identifying and isolating the source of exogenous changes in the parent firm's income. To address this, Dharmapala and Riedel (2013) adopt a method developed by Bertrand et al. (2002). They construct an expected earnings shock variable based on the earnings of firms operating in the same industry and country as the parent firm. This provides a measure of the parent's exogenous income before taxes and profit-shifting activities occur, facilitating the analysis of profit-shifting to foreign low-tax affiliates.

Finally, a novel development of the Hines–Rice approach in the accounting literature is exemplified by the work of Dyreng and Markle (2016).[8] Their method centers around the notion that the allocation of sales by a US-based multinational corporation between domestic and foreign customers is relatively non-manipulable due to the fixed location of final consumers. Building upon this premise, they contend that it becomes feasible to directly estimate the magnitude and direction of income shifting by analyzing discrepancies between US MNCs' geographical distribution of sales and their reported earnings. This approach avoids assuming that

[8] We refer the reader to other relevant papers in the accounting literature such as De Simone (2016), De Simone et al. (2022), Drake et al. (2022), Joshi (2020), and McGuire et al. (2019).

accounting rates of return would be equalized between US and foreign operations in the absence of income shifting. Nevertheless, its effectiveness relies heavily on the assumption that the location of sales is not susceptible to manipulation and remains unaffected by income shifting strategies.

## 2.6 Determinants of Tax Avoidance

In analyzing the theory underlying corporate tax avoidance, we have identified several contributions that have proposed models to examine the phenomenon. These contributions represent the logical starting point for developing measures to evaluate tax avoidance strategies, which researchers have used to investigate the factors determining tax avoidance. We now explain and categorize such factors or determinants following the conceptual framework proposed by Wilde and Wilson (2018), according to which the expected marginal benefits of tax avoidance should be evaluated against multiple marginal costs, as indicated in Fig. 2.2. Hence, holding benefits constant, differences in tax avoidance can be attributed to three types of costs: agency costs, implementation costs, and outcome costs after the implementation of tax planning strategies.

### *2.6.1 Agency Costs*

The first category of determinants of tax planning relates to the alignment or misalignment between ownership and management, also known as agency costs. The extent to which the interests of owners and managers are aligned has significant implications for operational and financial decisions. Research in tax accounting has extensively investigated the relationship between agency costs and corporate tax avoidance. Within this category, various subcategories have been explored, including the examination of tax planning and ownership structures (Mills & Newberry, 2001; Chen et al., 2010; Badertscher et al., 2013; McGuire et al., 2014), analysis of compensation arrangements (Phillips, 2003; Robinson et al., 2010; Armstrong et al., 2012; Gaertner, 2014; Powers et al., 2016; Chi et al., 2017), exploration of corporate governance features (Armstrong et al., 2015; Bird & Karolyi, 2017), and examination of individual executive characteristics (Dyreng et al., 2010; Chyz, 2013; Francis et al., 2014; Christensen et al., 2015; Feller & Schanz, 2017; Koester et al., 2017). Overall, such studies indicate that factors like corporate governance, ownership structure, executive characteristics, and compensation incentives collectively influence and either promote or discourage managers from making specific tax planning choices. Consequently, as managers adapt to unique ownership structures, incentives, and governance mechanisms, firms tend to display varying tax avoidance behaviors based on their specific incentives and governance characteristics. When a firm's governance mechanisms are ineffective or managerial motivations do not align with shareholder interests, managers may opt for sub-optimal decisions from the shareholders' viewpoint.

### 2.6.2 Implementation Costs

The second category in the WW framework focuses on variations in tax avoidance arising from implementation costs, which encompass the expenses associated with the execution and upkeep of tax planning strategies. Without agency considerations, properly incentivized managers should make tax-efficient corporate tax decisions when the marginal benefits outweigh the marginal costs. Research in this context explores the relationship between tax avoidance and various firm characteristics that influence the costs and benefits of tax planning activities, such as firm size (Zimmerman, 1983), planning costs (Mills et al., 1998), asset structure (Klassen & Laplante, 2012), capital structure (Graham & Tucker, 2006; De Vito & Jacob, 2023), financing constraints (Law & Mills, 2015; Edwards et al., 2016), internal controls (De Simone et al., 2015; Bauer, 2016), business strategy (Higgins et al., 2015), tax havens (Dyreng & Lindsey, 2009; Dyreng et al., 2015), and global operations (Rego, 2003). Additionally, studies examine how different operating environments impact tax avoidance opportunities and activities (Atwood et al., 2012; Kubick et al., 2015; De Simone, 2016). It is key to recognize that implementation costs are dynamic and can undergo changes due to evolving firm characteristics, business practices, global opportunities, competition, and alterations in tax regulations.

### 2.6.3 Outcome Costs

The final category of tax avoidance determinants focuses on studying factors associated with the expected outcome costs of tax planning strategies. This includes research on how regulatory monitoring (Hoopes et al., 2012; Kubick et al., 2016), regulation (Donohoe & McGill, 2011; Towery, 2017), external and internal monitors (Klassen et al., 2016), and external perceptions of tax planning activities, such as reputational costs, influence firms' tax-planning decisions (Gallemore et al., 2014; Brown et al., 2015; Chyz et al., 2013; Wilde, 2017).[9] Aligned with the conceptual framework proposed by WW, we contend that managers take into account agency

[9] Within the outcome costs category, another interesting line of research examines the association between corporate tax avoidance and social responsibility (CSR) activities (Lanis & Richardson, 2012; Hoi et al., 2013; Lanis & Richardson, 2015; Watson, 2015; Davis et al., 2016; Col & Patel, 2019). This stream of research generally finds mixed evidence on the link between certain forms of tax planning and CSR activities, which warrants further scrutiny in light of the current environmental, social, and governance (ESG) debate.

costs, implementation costs, and the anticipated outcome costs post-implementation when making decisions regarding the adoption of specific tax avoidance strategies.

To summarize, the three categories of costs—agency costs, implementation costs, and ex-post expected outcome costs—play a significant role in influencing variations in tax avoidance strategies across firms, assuming consistent expected benefits. However, similar to the perspective put forth by Wilde and Wilson (2018), it is worth noting that existing research on determinants of corporate tax behavior tends to examine these categories in *isolation*. This is partly due to limitations in research designs and the absence of a comprehensive theory that effectively explains the interconnectedness among these determinants (Brühne & Jacob, 2019). Therefore, we emphasize the need for further research that reconciles theory with empirical results and utilizes robust research designs capable of isolating the effects and their corresponding channels.

## 2.7 Consequences of Tax Avoidance

As discussed until now, firms, on average, engage in corporate tax avoidance. However, contrary to the prevailing view that firms always benefit from tax avoidance, the empirical evidence suggests a more nuanced picture, indicating that tax avoidance behavior can have substantial negative consequences for firm decision-making. Therefore, accounting scholars conducting tax avoidance research should focus on *whether* firms shelter their taxes and *what* happens to those firms that shelter taxes. In our view, this is essential for advancing the tax avoidance literature. In this context, we focus on the economic consequences of three key constructs that are particularly relevant to accounting scholars: financial reporting transparency, cost of capital, and firm value. In doing so, we employ the SW theoretical framework and systematically evaluate whether the marginal benefits of tax avoidance outweigh the marginal costs.[10]

### *2.7.1 Financial Reporting Transparency*

The link between tax avoidance and corporate transparency is motivated by the general thesis that tax avoidance requires complexity and obfuscation of information to prevent detection by the tax authority (Desai, 2005). Specifically, the extant

[10] It is important to note that, while our focus is on financial reporting transparency, cost of capital, and firm value, since they traditionally fall under the accounting domain, there could be other consequences of tax avoidance worth investigating. For example, Jacob (2022) argues that, beyond a few exceptions (Shevlin et al., 2019; Li et al., 2021), the role of tax avoidance in shaping the tax effects on investments is not yet fully understood.

literature suggests that tax avoidance decreases financial reporting transparency with undesirable effects on firms and investors for at least two reasons.

First, managers may engage in intra-company transactions through tax haven subsidiaries, which are not usually reported. This lack of transparency could allow managers to engage in activities that harm shareholders. Second, managers may make financial statements less readable by obfuscating financial information to the tax authority. This could mask a firm's actual operating performance and alter the perception of its value to investors. Taken as a whole, these theoretical arguments seem to support the notion that tax avoidance reduces financial reporting transparency, and the benefits of avoiding taxes (i.e., the tax savings from tax avoidance) do not outweigh the costs of reduced corporate transparency. Hence, empirically, there seems to be a *negative* association between tax avoidance and financial reporting transparency.

Notably, the results of the studies are not sensitive to alternative financial reporting and disclosure proxies. This is reassuring and further reinforces the idea that there is a negative association between tax avoidance and corporate transparency. This association is not simply a statistical artifact, but rather it is economically meaningful. For instance, in a sample of US firms, Kim et al. (2011) use stock price crash events as an indirect proxy for corporate transparency. They find that corporate tax avoidance positively correlates with a firm's stock price crash risk. This is because tax avoidance can facilitate managerial rent extraction and incentivize managers to hoard bad news. When the hidden bad news of sheltering taxes crosses a tipping point and comes out, investors punish the firm by selling its stocks. Chung et al. (2019) and Balakrishnan et al. (2019) reach similar conclusions. They argue that managerial opportunism, insider trading, and analyst pre-tax forecast errors are significantly higher for tax-avoiding firms with an opaque information environment. Kerr (2019) studies the association between financial reporting transparency and tax avoidance in a cross-country sample and through aggregate- and firm-level tests. He finds that countries and firms with greater levels of transparency exhibit lower levels of tax avoidance. Interestingly, he also finds that the effect of country-level transparency (i.e., adopting IFRS and enforcing insider trading laws) is incremental to firm-level transparency.

A few studies have examined how corporate tax avoidance affects specific information channels. Ayers et al. (2009) examine whether tax avoidance impairs or enhances the information content of taxable income relative to book income. They find that taxable income is less informative about firm performance for tax-avoiding firms, which is consistent with the notion that tax avoidance can obfuscate financial information. Donohoe and Robert Knechel (2014) focus on audit fees and find that tax-avoiding firms pay significantly higher audit fees. This is consistent with the notion that tax avoidance can make it more difficult for auditors to assess and understand a firm's financial performance. Chen et al. (2018) show that cross-border profit shifting limits investors' ability to understand the geographic location of the book and taxable income generation. This is consistent with the findings of Durnev et al. (2017) who find lower financial reporting quality for firms with operations in offshore financial centers. Joshi (2020) documents that mandating firms to disclose the geographic breakdown of their profitability and activities to the tax authority reduces tax-motivated income shifting. This suggests that providing more information to the tax authority can help to mitigate the negative consequences of tax avoidance. Finally, Inger et al. (2018) and Nguyen (2021) examine the financial statement readability of tax-avoiding firms. They find that tax avoidance reduces readability and makes firms less transparent. This suggests that managers of tax-avoiding firms deliberately conceal information from the tax authority, despite depriving shareholders of useful information.

The empirical evidence supports the theoretical prediction that tax avoidance reduces transparency in financial reporting. This is because tax avoidance often requires complexity and obfuscation of information, making it difficult for investors to assess the true financial performance of a firm. Additionally, tax avoidance can lower financial reporting quality, affecting a firm's cost of capital (Leuz & Verrecchia, 2005). This is because firms with lower financial reporting quality are more likely to be perceived as riskier by investors, which can then lead to higher costs of capital (Barth et al., 2013). In the next section, we discuss the literature on the association between tax avoidance and the cost of capital.

### 2.7.2 *Cost of Capital*

The relationship between tax avoidance and cost of capital is not straightforward and depends on whether the firm finances itself with equity or debt. Following the model of Lambert et al. (2007), tax avoidance could affect the cost of equity capital through two channels. The first channel is the cash flow channel. Suppose tax avoidance increases a firm's after-tax cash flows without simultaneously changing its covariance with other firms' cash flows. In that case, the model predicts that the cost of equity capital should decrease. The second channel is the covariance channel. When a firm shelters taxes, it likely does so in a way that alters its operations and fundamentals. By doing so, the firm could simultaneously alter the covariance of its cash flows with other firms' cash flows in the market. The sign of the covariance could be either positive or negative, depending on whether other firms in the market implement similar tax avoidance strategies. Hence, ex ante, it is difficult to predict whether the combination of the cash flow channel with the covariance channel makes the net effect of tax avoidance on the cost of equity capital positive or negative.

Early research on the relationship between tax avoidance and the cost of equity capital has produced mixed results. Goh et al. (2016) found that tax avoidance is associated with a lower cost of equity capital for US firms. In contrast, Brooks et al. (2016) found that tax avoidance is associated with an increase in the cost of equity capital for UK firms. Cook et al. (2017) found that the relationship between tax avoidance and the cost of equity capital depends on whether firms meet or exceed investors' expectations of tax avoidance.[11] Heitzman and Ogneva (2019) found that all firms in an industry may face a tax risk premium, regardless of whether they engage in tax-sheltering activities. These mixed results suggest that the relationship between tax avoidance and the cost of equity capital is complex and depends on several factors. We call for more research to reconcile theory and empirical results.

Regarding the relationship between tax avoidance and the cost of debt, both theoretical frameworks and empirical findings consistently indicate that lenders tend to disfavor tax avoidance activities. Unlike shareholders, debt holders face substantial downside risks and have asymmetric payoffs. Consequently, even if tax avoidance leads to tax savings, these benefits are likely to be realized primarily by shareholders through increased dividends. Moreover, the associated risks of tax avoidance often outweigh the potential rewards, and lenders may bear the consequences without sharing in the benefits, given their fixed claimant position. Thus, these arguments seem to support the notion of lenders expressing a negative stance toward tax avoidance.

Empirical research further substantiates this perspective. Hasan et al. (2014) were the first to investigate the relationship between tax avoidance and the cost of bank loans in the USA, finding that firms with higher levels of tax avoidance tend to incur higher spreads when securing bank loans. Similarly, Isin (2018) demonstrates a positive correlation between tax avoidance and loan spreads. However, the author also acknowledges that firms may have some flexibility in mitigating non-tax costs associated with tax avoidance premiums, highlighting the importance of considering the overall costs of tax avoidance as suggested by the SW framework. Moreover, Saavedra (2019) focuses on the riskier end of the continuum and observes that lenders incorporate tax risk into the pricing of syndicated loans. These findings align with Shevlin et al. (2020), who reveal that higher levels of tax avoidance contribute to increased bond yields and bank loan spreads due to the negative impact of tax risk on future pre-tax cash flows and volatility. In line with the risk argument, Platikanova (2017) adds that firms engaging in tax avoidance tend to have shorter debt maturities. However, it is worth noting that Lim (2011) presents contrasting results, showing that tax avoidance reduces the cost of debt. Nevertheless, these findings are based on a relatively limited sample of Korean firms and may not represent the wider population of firms.

[11] In Cook et al. (2017), the authors adopt two strategies to define the firm-level expected tax avoidance. In the first approach, they estimate the industry-year median for each tax avoidance measure and then take the first difference between each observation's reported tax avoidance and the industry-year median. The second approach involves using a 5-year auto-regressive model that looks at the realized tax avoidance to compute an expected value.

### 2.7.3 Firm Value

The empirical studies examined thus far provide evidence that tax avoidance can lead to significant negative consequences, including reduced financial reporting transparency and potential adverse effects on the cost of capital. However, knowledge about the potential impact on firm value is still limited in scope and primarily focuses on the United States. In theory, one could argue that tax avoidance should affect firm value due to its influence on the cost of capital. The intuition is that more transparent earnings and the associated information should positively impact the cost of capital, thereby increasing firm value. However, due to the theoretical ambiguities surrounding the cost of capital, it remains uncertain whether tax avoidance should enhance or diminish the firm value and shareholder wealth.

On the one hand, tax avoidance may positively affect firm value through the "cash flow channel," wherein higher future after-tax cash flows are anticipated. On the other hand, tax avoidance may also adversely affect firm value due to agency conflicts arising from reduced corporate transparency ( Desai & Dharmapala, 2009a). It is worth noting that successful tax avoidance requires obfuscation. Reduced transparency may allow managers and controlling shareholders to extract wealth and divert rents from minority shareholders (Desai et al., 2007). Consequently, tax avoidance may negatively impact firm value for poorly governed firms. These opposing views create an interesting empirical question regarding the net effect of tax avoidance on firm value.

For instance, McGuire et al. (2016) examine investors' evaluations of tax loss carryforwards (TLCFs) and find that firms with high levels of tax avoidance are positively valued. This is attributed to the signal that such firms can generate future taxable income to offset TLCFs through tax avoidance, thereby increasing firm value. Conversely, Chow et al. (2016) study the effect of tax avoidance on firm value in a setting where targets must disclose their tax avoidance activities. The authors discover that targets not engaged in tax-sheltering transactions receive higher takeover premiums, suggesting that acquirers are concerned about potential future liabilities associated with tax sheltering and consequently reward investors in non-tax-sheltering firms.

In summary, whether tax avoidance enhances or diminishes firm value remains an ongoing empirical question deserving further attention. Achieving a better understanding of the empirical relationship between tax avoidance and firm value is challenging, particularly in establishing causality, but it holds great relevance for advancing the accounting literature. We advocate for future research to contribute to this debate. Furthermore, efforts should be made to theoretically and empirically link the effects of tax avoidance and tax risk on firm value and the cost of capital. A notable example in this domain is the work of Jacob and Schütt (2020), who propose considering not only the level of tax payments (tax avoidance) but also the volatility of tax payments (tax risk) concerning firm value. We concur with their proposition, as the literature would greatly benefit from disentangling the firm value consequences of tax avoidance and tax risk by offering clear definitions and distinguishing between these concepts theoretically and empirically.

## 2.8 Summary

This chapter has provided a comprehensive overview of the theoretical foundations of tax avoidance. We have explored the various dimensions of tax avoidance and examined the factors contributing to its occurrence. While the existing literature has made significant progress in understanding these factors, it is important to acknowledge that there is still ample room for further refinement. Future research endeavors should address areas where theory lacks empirical evidence, or theoretical ambiguity persists while aligning with the direction suggested by empirical results.

Moreover, this chapter has underscored the profound impact of corporate tax avoidance on firm decisions. The consequences of tax avoidance extend beyond financial implications, encompassing critical aspects such as corporate transparency, the cost of capital, and firm value. Understanding corporate tax avoidance's determinants and broader ramifications is crucial for policymakers, tax authorities, and practitioners alike. By delving deeper into these consequences, future research can shed more light on the complex dynamics between tax avoidance and its wide-ranging effects, facilitating informed decision-making and policy formulation.

As researchers further explore tax avoidance behaviors, adopting more robust research designs that isolate the effects and respective channels involved is crucial. By employing rigorous methodologies, researchers can enhance the validity and reliability of their findings, thereby strengthening the overall knowledge of the phenomenon.

## References


Alexander, A., & De Vito, A. (2021). Reporting, financial, and firm value consequences of avoiding taxes. *Rivista Italiana di Ragioneria e di Economia Aziendale, 5*, 334–352.

Allingham, M. G., & Sandmo, A. (1972). Income tax evasion: A theoretical analysis. *Taxation: Critical Perspectives on the World Economy, 3*(1), 323–338.

Armstrong, C. S., Blouin, J. L., Jagolinzer, A. D., & Larcker, D. F. (2015). Corporate governance, incentives, and tax avoidance. *Journal of Accounting and Economics, 60*(1), 1–17.

Armstrong, C. S., Blouin, J. L., & Larcker, D. F. (2012). The incentives for tax planning. *Journal of Accounting and Economics, 53*(1–2), 391–411.

Atwood, T., Drake, M. S., Myers, J. N., & Myers, L. A. (2012). Home country tax system characteristics and corporate tax avoidance: International evidence. *The Accounting Review, 87*(6), 1831–1860.

Ayers, B. C., Jiang, J., & Laplante, S. K. (2009). Taxable income as a performance measure: The effects of tax planning and earnings quality. *Contemporary Accounting Research, 26*(1), 15–54.

Badertscher, B. A., Katz, S. P., & Rego, S. O. (2013). The separation of ownership and control and corporate tax avoidance. *Journal of Accounting and Economics, 56*(2–3), 228–250.

Badertscher, B. A., Katz, S. P., Rego, S. O., & Wilson, R. J. (2019). Conforming tax avoidance and capital market pressure. *The Accounting Review, 94*(6), 1–30.

Balakrishnan, K., Blouin, J. L., & Guay, W. R. (2019). Tax aggressiveness and corporate transparency. *The Accounting Review, 94*(1), 45–69.

Barth, M. E., Konchitchki, Y., & Landsman, W. R. (2013). Cost of capital and earnings transparency. *Journal of Accounting and Economics, 55*(2–3), 206–224.

Bauer, A. M. (2016). Tax avoidance and the implications of weak internal controls. *Contemporary Accounting Research, 33*(2), 449–486.

Beasley, M. S., Goldman, N. C., Lewellen, C. M., McAllister, M. (2021). Board risk oversight and corporate tax-planning practices. *Journal of Management Accounting Research, 33*(1), 7–32.

Becker, G. S. (1968). Crime and punishment: An economic approach. *Journal of Political Economy, 76*(2), 169–217.

Bertrand, M., Mehta, P., & Mullainathan, S. (2002). Ferreting out tunneling: An application to Indian business groups. *The Quarterly Journal of Economics, 117*(1), 121–148.

Bird, A., & Karolyi, S. A. (2017). Governance and taxes: Evidence from regression discontinuity (retracted). *The Accounting Review, 92*(1), 29–50.

Blouin, J. (2014). Defining and measuring tax planning aggressiveness. *National Tax Journal, 67*(4), 875–899.

Bonacchi, M., Marra, A., & Zarowin, P. (2019). Organizational structure and earnings quality of private and public firms. *Review of Accounting Studies, 24*, 1066–1113.

Brooks, C., Godfrey, C., Hillenbrand, C., & Money, K. (2016). Do investors care about corporate taxes? *Journal of Corporate Finance, 38*, 218–248.

Brown, J. L., Drake, K., & Wellman, L. (2015). The benefits of a relational approach to corporate political activity: Evidence from political contributions to tax policymakers. *The Journal of the American Taxation Association, 37*(1), 69–102.

Brühne, A., & Jacob, M. (2019). Corporate tax avoidance and the real effects of taxation: A review. *TRR 266 Accounting for Transparency Working Paper Series No. 34.*

Burgstahler, D. C., Hail, L., & Leuz, C. (2006). The importance of reporting incentives: Earnings management in European private and public firms. *The Accounting Review, 81*(5), 983–1016.

Chen, C.-W., Hepfer, B. F., Quinn, P. J., & Wilson, R. J. (2018). The effect of tax-motivated income shifting on information asymmetry. *Review of Accounting Studies, 23*, 958–1004.

Chen, K.-P., & Chu, C. C. (2005). Internal control versus external manipulation: A model of corporate income tax evasion. *Rand Journal of Economics, 36*(1), 151–164.

Chen, S., Chen, X., Cheng, Q., & Shevlin, T. (2010). Are family firms more tax aggressive than non-family firms? *Journal of Financial Economics, 95*(1), 41–61.

Chi, S., Huang, S. X., & Sanchez, J. M. (2017). CEO inside debt incentives and corporate tax sheltering. *Journal of Accounting Research, 55*(4), 837–876.

Chow, T., Klassen, K. J., & Liu, Y. (2016). Targets' tax shelter participation and takeover premiums. *Contemporary Accounting Research, 33*(4), 1440–1472.

Christensen, D. M., Dhaliwal, D. S., Boivie, S., & Graffin, S. D. (2015). Top management conservatism and corporate risk strategies: Evidence from managers' personal political orientation and corporate tax avoidance. *Strategic Management Journal, 36*(12), 1918–1938.

Chung, S. G., Goh, B. W., Lee, J., & Shevlin, T. (2019). Corporate tax aggressiveness and insider trading. *Contemporary Accounting Research, 36*(1), 230–258.

Chyz, J. A. (2013). Personally tax aggressive executives and corporate tax sheltering. *Journal of Accounting and Economics, 56*(2–3), 311–328.

Chyz, J. A., Leung, W. S. C., Li, O. Z., & Rui, O. M. (2013). Labor unions and tax aggressiveness. *Journal of Financial Economics, 108*(3), 675–698.

Coase, R. (1937). The nature of the firm. *Economica, 4*(16), 386–405.

Col, B., & Patel, S. (2019). Going to haven? corporate social responsibility and tax avoidance. *Journal of Business Ethics, 154*, 1033–1050.

Cook, K. A., Moser, W. J., & Omer, T. C. (2017). Tax avoidance and ex ante cost of capital. *Journal of Business Finance & Accounting, 44*(7–8), 1109–1136.

Crocker, K. J., & Slemrod, J. (2005). Corporate tax evasion with agency costs. *Journal of Public Economics, 89*(9–10), 1593–1610.

Davis, A. K., Guenther, D. A., Krull, L. K., & Williams, B. M. (2016). Do socially responsible firms pay more taxes? *The Accounting Review, 91*(1), 47–68.

De Simone, L. (2016). Does a common set of accounting standards affect tax-motivated income shifting for multinational firms? *Journal of Accounting and Economics, 61*(1), 145–165.
De Simone, L., Ege, M. S., & Stomberg, B. (2015). Internal control quality: The role of auditor-provided tax services. *The Accounting Review, 90*(4), 1469–1496.
De Simone, L., Klassen, K. J., & Seidman, J. K. (2022). The effect of income-shifting aggressiveness on corporate investment. *Journal of Accounting and Economics, 74*(1), 101491.
De Vito A, & Jacob M. (2023). The role of creditor protection in lending and tax avoidance. *Journal of Financial and Quantitative Analysis, 58*(5), 2096–2130. https://doi.org/10.1017/S0022109022001144
Desai, M. A. (2003). The divergence between book income and tax income. *Tax Policy and the Economy, 17*, 169–206.
Desai, M. A. (2005). The degradation of reported corporate profits. *Journal of Economic Perspectives, 19*(4), 171–192.
Desai, M. A. (2009). The decentering of the global firm. *World Economy, 32*(9), 1271–1290.
Desai, M. A., & Dharmapala, D. (2006). Corporate tax avoidance and high-powered incentives. *Journal of Financial Economics, 79*(1), 145–179.
Desai, M. A., & Dharmapala, D. (2009a). Corporate tax avoidance and firm value. *The Review of Economics and Statistics, 91*(3), 537–546.
Desai, M. A., & Dharmapala, D. (2009b). Earnings management, corporate tax shelters, and book-tax alignment. *National Tax Journal, 62*(1), 169–186.
Desai, M. A., Dyck, A., & Zingales, L. (2007). Theft and taxes. *Journal of Financial Economics, 84*(3), 591–623.
Desai, M. A., Foley, C. F., & Hines, J. R., Jr. (2006). The demand for tax haven operations. *Journal of Public Economics, 90*(3), 513–531.
Dharmapala, D. (2014). What do we know about base erosion and profit shifting? a review of the empirical literature. *Fiscal Studies, 35*(4), 421–448.
Dharmapala, D., & Riedel, N. (2013). Earnings shocks and tax-motivated income-shifting: Evidence from European multinationals. *Journal of Public Economics, 97*, 95–107.
Donohoe, M. P., & McGill, G. A. (2011). The effects of increased book-tax difference tax return disclosures on firm valuation and behavior. *Journal of the American Taxation Association, 33*(2), 35–65.
Donohoe, M. P., McGill, G. A., & Outslay, E. (2014). Risky business: The prosopography of corporate tax planning. *National Tax Journal, 67*(4), 851–874.
Donohoe, M. P., & Robert Knechel, W. (2014). Does corporate tax aggressiveness influence audit pricing? *Contemporary Accounting Research, 31*(1), 284–308.
Drake, K. D., Goldman, N. C., & Murphy, F. (2022). Foreign employment, income shifting, and tax uncertainty. *The Accounting Review, 97*(2), 183–212.
Dunbar, A., Higgins, D. M., Phillips, J. D., & Plesko, G. A. (2010). What do measures of tax aggressiveness measure? In: *Proceedings. Annual Conference on Taxation and Minutes of the Annual Meeting of the National Tax Association* (Vol. 103, pp. 18–26). JSTOR.
Durnev, A., Li, T., & Magnan, M. (2017). Beyond tax avoidance: Offshore firms' institutional environment and financial reporting quality. *Journal of Business Finance & Accounting, 44*(5–6), 646–696.
Dyreng, S. D., Hanlon, M., & Maydew, E. L. (2008). Long-run corporate tax avoidance. *The Accounting Review, 83*(1), 61–82.
Dyreng, S. D., Hanlon, M., & Maydew, E. L. (2010). The effects of executives on corporate tax avoidance. *The Accounting Review, 85*(4), 1163–1189.
Dyreng, S. D., Hoopes, J. L., & Wilde, J. H. (2016). Public pressure and corporate tax behavior. *Journal of Accounting Research, 54*(1), 147–186.
Dyreng, S. D., & Lindsey, B. P. (2009). Using financial accounting data to examine the effect of foreign operations located in tax havens and other countries on us multinational firms' tax rates. *Journal of Accounting Research, 47*(5), 1283–1316.

Dyreng, S. D., Lindsey, B. P., Markle, K. S., & Shackelford, D. A. (2015). The effect of tax and nontax country characteristics on the global equity supply chains of us multinationals. *Journal of Accounting and Economics, 59*(2–3), 182–202.

Dyreng, S. D., & Markle, K. S. (2016). The effect of financial constraints on income shifting by us multinationals. *The Accounting Review, 91*(6), 1601–1627.

Edwards, A., Schwab, C., & Shevlin, T. (2016). Financial constraints and cash tax savings. *The Accounting Review, 91*(3), 859–881.

Feller, A., & Schanz, D. (2017). The three hurdles of tax planning: How business context, aims of tax planning, and tax manager power affect tax expense. *Contemporary Accounting Research, 34*(1), 494–524.

Francis, B. B., Hasan, I., Wu, Q., & Yan, M. (2014). Are female CFOs less tax aggressive? evidence from tax aggressiveness. *The Journal of the American Taxation Association, 36*(2), 171–202.

Frank, M. M., Lynch, L. J., & Rego, S. O. (2009). Tax reporting aggressiveness and its relation to aggressive financial reporting. *The Accounting Review, 84*(2), 467–496.

Frischmann, P. J., Shevlin, T., & Wilson, R. (2008). Economic consequences of increasing the conformity in accounting for uncertain tax benefits. *Journal of Accounting and Economics, 46*(2–3), 261–278.

Gaertner, F. B. (2014). CEO after-tax compensation incentives and corporate tax avoidance. *Contemporary Accounting Research, 31*(4), 1077–1102.

Gallemore, J., Maydew, E. L., Thornock, J. R. (2014). The reputational costs of tax avoidance. *Contemporary Accounting Research, 31*(4), 1103–1133.

Goh, B. W., Lee, J., Lim, C. Y., & Shevlin, T. (2016). The effect of corporate tax avoidance on the cost of equity. *The Accounting Review, 91*(6), 1647–1670.

Graham, J. R., & Tucker, A. L. (2006). Tax shelters and corporate debt policy. *Journal of Financial Economics, 81*(3), 563–594.

Grubert, H., & Mutti, J. (1991). Taxes, tariffs and transfer pricing in multinational corporate decision making. *Review of Economics and Statistics, 73*, 285–293. https://econpapers.repec.org/article/tprrestat/v_3a73_3ay_3a1991_3ai_3a2_3ap_3a285-93.htm

Hanlon, M. (2003). What can we infer about a firm's taxable income from its financial statements? *National Tax Journal, 56*(4), 831–863.

Hanlon, M., & Heitzman, S. (2010). A review of tax research. *Journal of Accounting and Economics, 50*(2), 127–178.

Hasan, I., Hoi, C. K. S., Wu, Q., & Zhang, H. (2014). Beauty is in the eye of the beholder: The effect of corporate tax avoidance on the cost of bank loans. *Journal of Financial Economics, 113*(1), 109–130.

Heitzman, S. M., & Ogneva, M. (2019). Industry tax planning and stock returns. *The Accounting Review, 94*(5), 219–246.

Henry, E., & Sansing, R. (2018). Corporate tax avoidance: Data truncation and loss firms. *Review of Accounting Studies, 23*, 1042–1070.

Higgins, D., Omer, T. C., & Phillips, J. D. (2015). The influence of a firm's business strategy on its tax aggressiveness. *Contemporary Accounting Research, 32*(2), 674–702.

Hines Jr, J. R., & Rice, E. M. (1994). Fiscal paradise: Foreign tax havens and American business. *The Quarterly Journal of Economics, 109*(1), 149–182.

Hoi, C. K., Wu, Q., & Zhang, H. (2013). Is corporate social responsibility (CSR) associated with tax avoidance? evidence from irresponsible CSR activities. *The Accounting Review, 88*(6), 2025–2059.

Hoopes, J. L., Mescall, D., & Pittman, J. A. (2012). Do IRS audits deter corporate tax avoidance? *The Accounting Review, 87*(5), 1603–1639.

Huizinga, H., & Laeven, L. (2008). International profit shifting within multinationals: A multi-country perspective. *Journal of Public Economics, 92*(5–6), 1164–1182.

Inger, K. K., Meckfessel, M. D., Zhou, M., & Fan, W. (2018). An examination of the impact of tax avoidance on the readability of tax footnotes. *The Journal of the American Taxation Association, 40*(1), 1–29.

Isin, A. A. (2018). Tax avoidance and cost of debt: The case for loan-specific risk mitigation and public debt financing. *Journal of Corporate Finance, 49*, 344–378.
Jacob, M. (2022). Real effects of corporate taxation: A review. *European Accounting Review, 31*(1), 269–296.
Jacob, M., & Schütt, H. H. (2020). Firm valuation and the uncertainty of future tax avoidance. *European Accounting Review, 29*(3), 409–435.
Jones, J. J. (1991). Earnings management during import relief investigations. *Journal of Accounting Research, 29*(2), 193–228.
Joshi, P. (2020). Does private country-by-country reporting deter tax avoidance and income shifting? evidence from BEPS action item 13. *Journal of Accounting Research, 58*(2), 333–381.
Kerr, J. N. (2019). Transparency, information shocks, and tax avoidance. *Contemporary Accounting Research, 36*(2), 1146–1183.
Kim, J.-B., Li, Y., & Zhang, L. (2011). Corporate tax avoidance and stock price crash risk: Firm-level analysis. *Journal of Financial Economics, 100*(3), 639–662.
Klassen, K. J., & Laplante, S. K. (2012). Are us multinational corporations becoming more aggressive income shifters? *Journal of Accounting Research, 50*(5), 1245–1285.
Klassen, K. J., Lisowsky, P., & Mescall, D. (2016). The role of auditors, non-auditors, and internal tax departments in corporate tax aggressiveness. *The Accounting Review, 91*(1), 179–205.
Koester, A., Shevlin, T., & Wangerin, D. (2017). The role of managerial ability in corporate tax avoidance. *Management Science, 63*(10), 3285–3310.
Kubick, T. R., Lynch, D. P., Mayberry, M. A., & Omer, T. C. (2015). Product market power and tax avoidance: Market leaders, mimicking strategies, and stock returns. *The Accounting Review, 90*(2), 675–702.
Kubick, T. R., Lynch, D. P., Mayberry, M. A., & Omer, T. C. (2016). The effects of regulatory scrutiny on tax avoidance: An examination of sec comment letters. *The Accounting Review, 91*(6), 1751–1780.
Lambert, R., Leuz, C., & Verrecchia, R. E. (2007). Accounting information, disclosure, and the cost of capital. *Journal of Accounting Research, 45*(2), 385–420.
Lanis, R., & Richardson, G. (2012). Corporate social responsibility and tax aggressiveness: An empirical analysis. *Journal of Accounting and Public Policy, 31*(1), 86–108.
Lanis, R., & Richardson, G. (2015). Is corporate social responsibility performance associated with tax avoidance? *Journal of Business Ethics, 127*, 439–457.
Law, K. K., & Mills, L. F. (2015). Taxes and financial constraints: Evidence from linguistic cues. *Journal of Accounting Research, 53*(4), 777–819.
Leuz, C., & Verrecchia, R. E. (2005). Firms' capital allocation choices, information quality, and the cost of capital. *Information Quality, and the Cost of Capital* (January 2005).
Li, Q., Ma, M. S., & Shevlin, T. (2021). The effect of tax avoidance crackdown on corporate innovation. *Journal of Accounting and Economics, 71*(2–3), 101382.
Lim, Y. (2011). Tax avoidance, cost of debt and shareholder activism: Evidence from Korea. *Journal of Banking & Finance, 35*(2), 456–470.
McClure, C. G. (2023). How costly is tax avoidance? evidence from structural estimation. *The Accounting Review, 98*(6), 353–380. https://doi.org/10.2308/TAR-2020-0465
McGill, G. A., & Outslay, E. (2004). Lost in translation: Detecting tax shelter activity in financial statements. *National Tax Journal, 57*(3), 739–756.
McGuire, S. T., Wang, D., & Wilson, R. J. (2014). Dual class ownership and tax avoidance. *The Accounting Review, 89*(4), 1487–1516.
McGuire, S. T., Neuman, S. S., Olson, A. J., & Omer, T. C. (2016). Do investors use prior tax avoidance when pricing tax loss carryforwards? *The Journal of the American Taxation Association, 38*(2), 27–49.
McGuire, S. T., Rane, S. G., & Weaver, C. D. (2019). Cost structure and tax-motivated income shifting. *The Accounting Review, 98*(7), 1–22.
Menicacci, L. (2022a). *Book-Tax conformity in the IFRS Era: Evidence from Italian Listed Companies*. FrancoAngeli.

Menicacci, L. (2022b). Financial reporting and book-tax conformity: A review of the issues. *Financial Reporting, (2022/1)*, 41–77.
Mills, L., Erickson, M. M., & Maydew, E. L. (1998). Investments in tax planning. *The Journal of the American Taxation Association, 20*(1), 1.
Mills, L. F. (1998). Book-tax differences and internal revenue service adjustments. *Journal of Accounting Research, 36*(2), 343–356.
Mills, L. F., & Newberry, K. J. (2001). The influence of tax and nontax costs on book-tax reporting differences: Public and private firms. *Journal of the American Taxation Association, 23*(1), 1–19.
Nguyen, J. H. (2021). Tax avoidance and financial statement readability. *European Accounting Review, 30*(5), 1043–1066.
OECD (2018). *OECD/G20 inclusive framework on BEPS: Progress report July 2018–May 2019*. Technical report, Organisation for Economic Co-operation and Development.
Phillips, J. D. (2003). Corporate tax-planning effectiveness: The role of compensation-based incentives. *The Accounting Review, 78*(3), 847–874.
Platikanova, P. (2017). Debt maturity and tax avoidance. *European Accounting Review, 26*(1), 97–124.
Powers, K., Robinson, J. R., & Stomberg, B. (2016). How do CEO incentives affect corporate tax planning and financial reporting of income taxes? *Review of Accounting Studies, 21*, 672–710.
Rego, S. O. (2003). Tax-avoidance activities of us multinational corporations. *Contemporary Accounting Research, 20*(4), 805–833.
Robinson, J. R., Sikes, S. A., & Weaver, C. D. (2010). Performance measurement of corporate tax departments. *The Accounting Review, 85*(3), 1035–1064.
Saavedra, D. (2019). Is tax volatility priced by lenders in the syndicated loan market? *European Accounting Review, 28*(4), 767–789.
Scholes, M. S., Wolfson, M. A., Erickson, M., Maydew, E., & Shevlin, T. (2014). *Taxes & business strategy*. Prentice Hall.
Schwab, C. M., Stomberg, B., & Williams, B. M. (2022). Effective tax planning. *The Accounting Review, 97*(1), 413–437.
Shackelford, D. A., & Shevlin, T. (2001). Empirical tax research in accounting. *Journal of Accounting and Economics, 31*(1–3), 321–387.
Shevlin, T. (2001). Corporate tax shelters and book-tax differences. *Tax Law Review, 55*, 427.
Shevlin, T., Shivakumar, L., & Urcan, O. (2019). Macroeconomic effects of corporate tax policy. *Journal of Accounting and Economics, 68*(1), 101233.
Shevlin, T., Urcan, O., & Vasvari, F. P. (2020). Corporate tax avoidance and debt costs. *The Journal of the American Taxation Association, 42*(2), 117–143.
Slemrod, J. (2004). The economics of corporate tax selfishness. *National Tax Journal, 57*(4), 877–899.
Slemrod, J., & Yitzhaki, S. (2002). Tax avoidance, evasion, and administration. In: *Handbook of public economics* (Vol. 3, pp. 1423–1470). Elsevier.
Towery, E. M. (2017). Unintended consequences of linking tax return disclosures to financial reporting for income taxes: Evidence from schedule UTP. *The Accounting Review, 92*(5), 201–226.
Watson, L. (2015). Corporate social responsibility, tax avoidance, and earnings performance. *The Journal of the American Taxation Association, 37*(2), 1–21.
Wilde, J. H. (2017). The deterrent effect of employee whistleblowing on firms' financial misreporting and tax aggressiveness. *The Accounting Review, 92*(5), 247–280.
Wilde, J. H., & Wilson, R. J. (2018). Perspectives on corporate tax planning: Observations from the past decade. *Journal of the American Taxation Association, 40*(2), 63–81.
Wilson, R. J. (2009). An examination of corporate tax shelter participants. *The Accounting Review, 84*(3), 969–999.
Zimmerman, J. L. (1983). Taxes and firm size. *Journal of Accounting and Economics, 5*, 119–149.

# Chapter 3
# Network Analysis: A Mathematical Framework

## 3.1 Introduction to Graph Theory

One of the objectives of this monograph is to investigate the relationships between entities, as identified in our sample of papers that have studied tax avoidance. For this purpose, graph theory emerges as a reliable and relatively simple tool that enables understanding bonds across individuals (Borgatti et al., 2009). Graph theory is a branch of mathematics that studies the properties and applications of graphs, which are abstract structures that model pairwise relations between objects. Graphs can represent many real-world phenomena, such as networks, circuits, maps, games, puzzles, etc. A graph consists of a set of vertices (also called nodes or points) and a set of edges (also called links or lines) that connect some pairs of vertices. An edge is said to be incident on its endpoints, and two edges are said to be adjacent if they share a common endpoint. A vertex is said to be adjacent to another vertex if an edge connects them. The study of graphs has applications in various fields, including computer science, engineering, physics, biology, and social sciences. Graph theory has become increasingly popular in the field of network analysis, which involves the study of complex systems of interconnected components. Different types of graphs depend on the nature and the number of edges. A simple graph is a graph that has no loops (edges that connect a vertex to itself) and no multiple edges (more than one edge between the same pair of vertices). A multigraph is a graph that may have loops and multiple edges. A directed graph (or digraph) is a graph that assigns a direction to each edge, indicating the source and the target vertex of the edge. A weighted graph is a graph that assigns a numerical value (or weight) to each edge.

One of the basic concepts in graph theory is the degree of a vertex, which is the number of edges incident on it. For a directed graph, the degree of a vertex can be divided into the in-degree (the number of edges entering the vertex) and the out-degree (the number of edges leaving the vertex). The sum of the degrees of all vertices in a graph equals twice the number of edges. Another essential concept is the path in a graph, which is a sequence of vertices such that an edge connects each

 
A. De Vito, F. Grossetti, *Tax Avoidance Research*, SIDREA Series in Accounting and Business Administration, https://doi.org/10.1007/978-3-031-51765-5_3

pair of consecutive vertices. The length of a path is the number of edges in it. A path is simple if it does not repeat any vertex or edge. A cycle is a path that starts and ends at the same vertex and has a length of at least 3. A graph is said to be connected if there is a path between any two vertices in it. A connected component of a graph is a maximal connected sub-graph of it. Graph theory encompasses many concepts that give birth to variations, such as trees, planarity, coloring, matchings, network flows, extremal graph theory, and algebraic graph theory. In the following sections, we will cover the mathematical framework of graph theory to build a solid knowledge base that allows us to extract valuable insights through network analysis.

When examining an observed network and its structure, scholars are interested in its appearance and understanding its underlying origins. We conceive the network as a result of various processes associated with the complex system that captures our attention, and we aim to identify the essential aspects of these processes. Additionally, it is crucial to consider how the network was obtained, including the measurement and construction processes involved. These considerations motivate network modeling and the associated tools of statistical inference. Network modeling has garnered significant attention, broadly falling into two categories: mathematical and statistical models. Mathematical models are typically defined by simple probabilistic rules that generate a network, often aiming to capture specific mechanisms or principles, such as the "rich get richer" concept. On the other hand, statistical models, also often probabilistic, are designed, at least in part, to fit observed data. These models allow for evaluating the explanatory power of certain variables in forming network connections and employ formal principles of statistical inference for assessing the goodness of fit. While there is some overlap between these two classes of models, they are largely distinct in the relevant literature.

The simplest example of a mathematical network model involves randomly assigning edges to pairs of vertices based on independent and identically distributed coin tosses—a concept akin to the well-known Erdös–Rényi formulation of a random graph (Erdős et al., 1960). This model has been extensively studied since the 1960s. Its effectiveness stems from its ability to comprehensively understand its properties, such as the emergence of cohesive structures based on edge probabilities and its value as a benchmark for comparing more intricate models. In the literature, numerous statistical network models have been suggested, several exhibiting resemblances to established model classes pertaining to classical statistics. For instance, exponential random graph models (Lusher et al., 2013) resemble generalized linear models based on an exponential family form. Similarly, latent network models (Kolaczyk et al., 2020), which propose that edges may arise partially from unmeasured and potentially unknown variables, directly parallel the use of latent variables in hierarchical statistical modeling. Finally, stochastic block models (Holland et al., 1983; Abbe, 2017) can be viewed as a mixture model. However, it is important to note that the specification and fitting of such models are typically more complex, given the high-dimensional and dependent nature of the data.

## 3.2 What Is a Graph?

Before moving forward and discussing the properties of a network system, we define the fundamental concepts that are at the basis of such entities. A **graph** $G$ consists of two sets: a set $V(G)$ of **vertices** (or **nodes** or **points**) and a set $E(G)$ of **edges** (or **links** or **lines**) that connect some pairs of vertices. Formally, we denote a graph $G$ with vertex set $V$ and edge set $E$ as follows:

$$G = (V, E). \tag{3.1}$$

The simplest form of a graph is called *undirected simple graph*. We also write $|V|$ and $|E|$ to denote the number of vertices and edges in $G$, respectively. Alternatively, we write the same as follows:

$$V = v_1, v_2, \ldots, v_n, \tag{3.2}$$

$$E = (u, v) \mid u, v \in V. \tag{3.3}$$

An edge $e$ in $E$ can be represented by an unordered pair $\{u, v\}$ of vertices in $V$, where $u$ and $v$ are called the **endpoints** of $e$. We write $e = uv$ or $e = vu$ for simplicity. We say that $e$ is **incident** on $u$ and $v$ and that $u$ and $v$ are **adjacent** to each other.

In general, we recognize three classes of graphs:

- **Undirected Graph (UG)**. In an undirected graph, each edge is an unordered pair of vertices. In other words, the edge $(u, v)$ is equivalent to the edge $(v, u)$.
- **Directed Graph (DG)**. In a directed graph, each edge is an ordered pair of vertices. In other words, the edge $(u, v)$ represents a directed edge from vertex $u$ to vertex $v$.
- **Weighted Graph (WG)**. In a weighted graph, each edge is assigned a weight or cost. The weight can represent various properties, such as distance, time, or strength. The weight is represented by the function $w : E \rightarrow \mathbb{R}$, where $\mathbb{R}$ is the set of real numbers.

In other words, for DGs, we introduce a concept of "direction." This has direct implications as we formalize the flow of information (i.e., the information transport) by saying that moving from vertex $u$ to vertex $v$ is not the same as moving from $v$ to $u$. This is a rather important concept that will be extensively used in some of the analyses discussed in this chapter.

### *3.2.1 Properties of Graphs*

Graphs have several properties and characteristics that are useful in specific situations. This section discusses some of the most essential features representing fundamental building blocks in graph theory. These include connectivity, degree, diameter, and planarity, among others. We will also provide examples to illustrate their applications.

#### 3.2.1.1 Connectivity

Connectivity is a fundamental property in graph theory and network analysis that characterizes the ability of a graph or a network to maintain a connected structure. It quantifies the extent to which nodes in the graph are interconnected, enabling the flow of information, resources, or influence between them. In graph theory, connectivity can be defined in various ways depending on the specific context and requirements. In general, connectivity refers to the ability to traverse the graph from one vertex to another. A graph is said to be connected if there exists a path between any two vertices in the graph. Formally, a graph $G = (V, E)$ is connected if, for every pair of distinct vertices $u, v \in V$, there exists a path from $u$ to $v$. A path is a sequence of vertices connected by edges. For example, consider the following graph (Fig. 3.1):

This graph is connected because there is a path between any two vertices given by the path $A \rightarrow B \rightarrow C \rightarrow D \rightarrow E$. A graph can be partitioned into connected components if it is not connected. A connected component is a maximal connected sub-graph of the original graph. For example, the graph in Fig. 3.2 has two connected components given by $\{A, B, C, D, E\}$ and $\{F, G, H\}$.

Another commonly used measure is vertex connectivity, denoted by $\kappa(G)$, which represents the minimum number of nodes that must be removed from a graph $G$ to disconnect it. Mathematically, if $\kappa(G)$ is equal to $k$, it implies that $G$ is $k$-vertex-connected. A graph is said to be connected if it contains a path between every pair of vertices, and a disconnected graph can be partitioned into two or more connected components. Similarly, a related notion is edge connectivity, denoted by $\lambda(G)$, which is the minimum number of edges that must be removed from a graph $G$ to disconnect it. If $\lambda(G)$ is equal to $l$, the graph $G$ is considered $l$-edge-

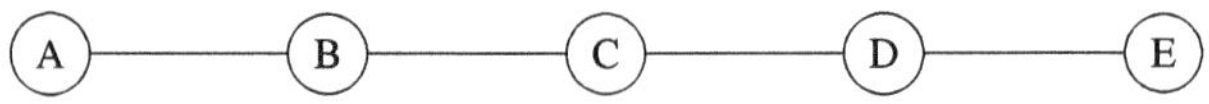


**Fig. 3.1** A simple connected graph

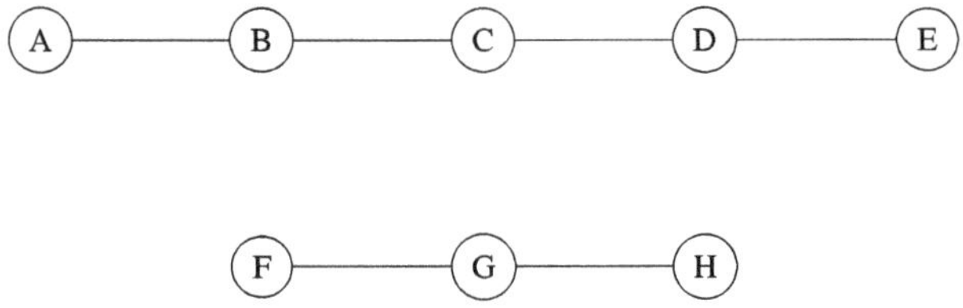


**Fig. 3.2** Two simple connected graphs

connected. The edge connectivity of a graph can be greater than or equal to its vertex connectivity, and both measures provide valuable insights into the robustness and resilience of networks. In addition to these global connectivity measures, local connectivity measures capture the connectivity properties of individual nodes. For instance, the degree of a node $v$ in a graph $G$, denoted by $\deg(v)$, represents the number of edges incident to $v$. We will discuss the degree property in the next section.

Connectivity is closely related to the concept of paths and connectivity between pairs of nodes. A path in a graph $G$ is a sequence of vertices $v_1, v_2, \ldots, v_n$ such that there is an edge between $v_i$ and $v_{i+1}$ for $1 \leq i \leq n - 1$. The length of a path is defined as the number of edges it contains. A path exists between any pair of nodes in a connected graph, while specific pairs of nodes are unreachable in a disconnected graph. Connectivity plays a crucial role in various real-world applications. In transportation networks, connectivity ensures that all locations are reachable, facilitating the efficient movement of people and goods. In social networks, connectivity allows for the diffusion of information, influence, and the construction of social ties. In communication networks, connectivity guarantees the reliable transmission of data across nodes. Robust connectivity is vital in power grids, ensuring uninterrupted electricity distribution and preventing cascading failures. Understanding and analyzing connectivity in graphs and networks can help identify critical nodes or edges that, when disrupted, can significantly impact the overall connectivity and functionality of the system. Connectivity measures provide insights into network resilience, vulnerability, and the potential spread of disruptions. They aid in designing more robust and efficient networks, improving their fault tolerance, and developing strategies for network optimization and resource allocation.

#### 3.2.1.2 Measures of Centrality

Another relevant concept in graph theory is centrality. This quantifies the importance or prominence of nodes within a network. It measures the extent to which a node occupies a *central* or *influential* position in the network structure. Centrality provides valuable insights into the relative significance of nodes and their impact on information flow, influence propagation, and network dynamics.

Several centrality measures are commonly used in graph theory, each capturing different aspects of node importance. The degree of centrality characterizes the connectivity of individual nodes in a graph or a network. The degree of a node represents the number of edges that are incident to that node. In mathematical notation, the degree of a node $v$ in a graph $G$ is denoted as $\deg(v)$. In other words, it is the number of edges connected to $v$. Consider the graph in Fig. 3.3. The vertex $A$ has degree 2 since it is incident to edges $\{AB, AD\}$. The vertex $B$ also has degree 2 since it is incident to edges $\{AB, BC\}$. The vertex $C$ has degree 2 since it is incident to edges $\{BC, CD\}$. Finally, the vertex $D$ has degree 2 since it is incident to edges $\{AD, CD\}$.

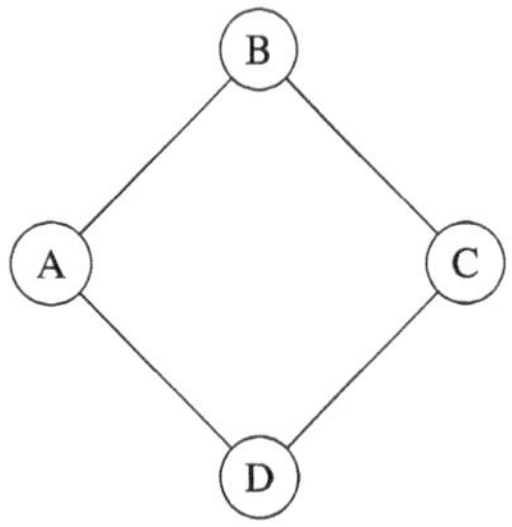


**Fig. 3.3** A four-node connected graph

The degree centrality provides important insights into the connectivity and importance of nodes within a network. Nodes with higher degrees typically have more connections to other nodes, indicating greater influence or centrality in the network. These nodes are crucial in information dissemination, resource flow, or influence propagation within the network. The degree distribution of a graph describes the probability distribution of node degrees across the entire network. It provides valuable information about the structural properties of the network. For example, in a random graph, the degree distribution follows a specific probability distribution, such as the Poisson distribution. In scale-free networks, however, the degree distribution follows a power-law distribution, indicating the presence of a few highly connected nodes or "hubs."[1]

Furthermore, the degree centrality is also relevant in various applications. In social networks, nodes with high degrees represent individuals with many social connections, such as popular influencers or well-connected individuals. In transportation networks, nodes with high degrees correspond to important hubs, such as airports or major intersections, enabling efficient transportation between different locations. In biological networks, the degree property can represent the number of interactions or connections between molecules or proteins, providing insights into biological processes. By extension, in the context of this study, the degree property sheds light on the connections among scholars, institutions, journals, and countries. Analyzing the degree of property helps in understanding network resilience, vulnerability, and the spread of information or influence within a network. Highly connected nodes are more susceptible to spreading or receiving information, making them potential targets for strategic interventions. Furthermore, the degree property is crucial in designing robust networks by identifying critical nodes whose removal may lead to network fragmentation or loss of connectivity.

Another commonly used centrality measure is betweenness centrality, which quantifies the extent to which a node lies on the shortest paths between other nodes in the network. Mathematically, the betweenness centrality $g$ of a node $v$ is given by

[1] The term "scale-free" refers to the property that the degree distribution does not have a characteristic scale or cutoff point.

$$g(v) = \sum_{s \neq v \neq t} \frac{\sigma_{st}(v)}{\sigma_{st}}, \tag{3.4}$$

where $\sigma_{st}$ is the total number of shortest paths between nodes $s$ and $t$, and $\sigma_{st}(v)$ is the number of those shortest paths that pass through node $v$. Nodes with higher betweenness centrality act as critical connectors or bridges, facilitating the flow of information or resources between different parts of the network.

Closeness centrality is another measure that quantifies how close a node is to other nodes in the network. It is based on the average shortest path distance between a node and all other nodes in the graph. Mathematically, the closeness centrality $C$ of a node $v$ is given by

$$C(v) = \frac{1}{\sum_{u \neq v} d(u, v)}, \tag{3.5}$$

where $d(u, v)$ is the shortest path between nodes $u$ and $c$. Nodes with higher closeness centrality are more central and can communicate or spread information more efficiently within the network.

Another relevant measure of centrality is the eigenvector centrality, also known as *eigencentrality*. This quantity takes into account both the centrality of a node and the centrality of its neighbors. It assigns higher centrality scores to nodes connected to other highly central nodes. The eigenvector centrality $\xi$ of a node in a graph can be mathematically defined as follows. Let $G = (V, E)$ be a graph with $V$ representing the set of nodes and $E$ representing the set of edges. To define $\xi$, we must introduce a new concept called *adjacency matrix*. The adjacency matrix is a square matrix used to represent the connectivity of a graph. Let $G = (V, E)$ be a graph with $n$ nodes, where $V$ represents the set of nodes and $E$ represents the set of edges. The adjacency matrix $A$ of the graph $G$ is an $n \times n$ matrix, where $A_{ij}$ is equal to 1 if there is an edge between nodes $i$ and $j$ and 0 otherwise.[2]

The eigenvector centrality of a node $v$ is defined as the $v$-th component of the principal eigenvector $\mathbf{x}$ of the adjacency matrix $A$. Mathematically, it can be represented as

$$\xi(v) = \frac{x_v}{\lambda}, \tag{3.6}$$

where $\lambda$ is a normalization factor.

To compute the eigenvector centrality, we start by initializing an initial centrality vector $\mathbf{x}^{(0)}$ with non-negative values. Then, we iteratively update $\mathbf{x}^{(t)}$ until conver-

[2] The adjacency matrix provides a compact representation of the graph's connectivity structure. By examining the matrix elements, we can determine which nodes are connected and visualize the relationships between different nodes in the graph. It is a fundamental tool for graph-related computations, such as finding paths, detecting cycles, and analyzing network properties.

gence using the power iteration method such as

$$\mathbf{x}^{(t+1)} = \frac{A\mathbf{x}^{(t)}}{\|A\mathbf{x}^{(t)}\|}, \tag{3.7}$$

where $t$ represents the iteration number and $\|\cdot\|$ denotes the Euclidean norm. The power iteration method finds the principal eigenvector of the adjacency matrix, which corresponds to the eigenvector with the largest eigenvalue. The normalization factor $\lambda$ ensures that the centrality values are scaled appropriately.[3]

The eigenvector centrality captures the idea that a node's centrality depends not only on its direct connections but also on the centrality of its neighbors. Nodes connected to highly central nodes will have higher eigenvector centrality values, contributing to their overall influence in the network. The eigenvector centrality measure provides insights into nodes' relative importance and influence in a network. Nodes with higher eigenvector centrality are considered more central and influential, as they are connected to other highly central nodes.

The last centrality measure that we use in this study is the authority score (Kleinberg, 1999). Such a measure identifies a node's influence by focusing on identifying authoritative nodes in a network, which are considered reliable and trustworthy sources of information. In a directed graph, where edges have a specific direction from one node to another, the authority score is often used with the hub score. These measures are key components of the HITS (Hypertext-Induced Topic Selection) algorithm proposed by Jon Kleinberg (Kleinberg, 1999).

Let us consider a directed graph represented by $G(V, E)$, where $V$ represents the set of nodes (vertices) and $E$ represents the set of edges. The authority score of a node $i$ is denoted as $A(i)$. The authority score of a node is influenced by the number and quality of incoming links it receives from other authoritative nodes in the network. The authority score can be calculated iteratively using Algorithm 1.

**Algorithm 1** Authority Score Calculation

1: Initialize the authority score for all nodes to 1, i.e., $A(i) = 1$ for all $i \in V$.
2: Iterate until convergence:
3: **repeat**
4: Update the hub score for each node $j$ as the sum of authority scores of nodes linking to $j$: $H(j) = \sum A(i)$, where $(i, j)$ is an incoming edge to node $j$.
5: Update the authority score for each node $i$ as the sum of hub scores of nodes linked from $i$: $A(i) = \sum H(j)$, where $(i, j)$ is an outgoing edge from node $i$.
6: Normalize the authority scores by dividing each score by the sum of all authority scores: $A(i) = \frac{A(i)}{\sum A(k)}$, where $k$ iterates over all nodes in $V$.
7: **until** convergence.

[3] It is important to note that the power iteration method may converge to a local eigenvector rather than the principal eigenvector in some cases, especially for disconnected graphs. In such cases, additional techniques like the personalized PageRank algorithm can be employed to compute the eigenvector centrality.

This iterative process continues until the authority scores converge to a stable state. The convergence indicates that the authority scores have reached a point where further iterations do not significantly change the scores. The authority score is instrumental in information retrieval systems and search engines, where it helps identify reliable sources of information. Nodes with higher authority scores are considered more trustworthy and credible, and their content is given more weight in ranking search results. While the authority score provides valuable insights into node importance, it is important to note that its effectiveness depends on the network structure and the context in which it is applied.

## 3.3 Building a Network

The initial step we made was to build networks to understand all the different links between the actors. Networks are based on connections between papers in the corpus and their references. Therefore, we extracted the references for the 612 papers in our corpus, leveraging the Scopus API function. Figure 3.4 shows the code used to run the extraction query, included in the R package `rscopus` (Muschelli, 2019).

This extraction query results in a list of 7,650 references associated with our dataset. These references include all the citations made in the corpus papers without distinction regarding reference type, journal, and publication date.

### *3.3.1 Data Cleaning*

Before proceeding with the network analysis baseline dataset, we cleaned all the references retrieved from the previously illustrated query, applying the same eligibility criteria discussed in Sect. 4.2.1, except for the publishing date. That is, the cited papers can have a publishing date also earlier than January 1, 2000. First, all the references without titles, authors, affiliations, or publishers have been removed.

```
citing_articles <-
  tax_avoidance_corpus_nodup$`dc:identifier` # extracting the IDs of our articles

citation_list <- vector("list", length(docvec))

for (i in seq_along(citation_list)) {
  citation_list[[i]] <- abstract_retrieval(
    citing_articles[i],
    identifier = "scopus_id",
    headers = insttoken,
    view = "FULL"
  )
}
```

**Fig. 3.4** The query to Scopus Search API executed with the function `abstract_retrieval()` from the package `rscopus` to extract reference details

Second, the same filter for publishing journals has been applied for the references to keep the analysis representative of the current situation of the tax avoidance literature without polluting it with lower quality scientific production. The result is a final reference list of 6,861 studies cited by our corpus.

### 3.3.2 Nodes and Edges Alignment

Edge alignment refers to the arrangement or positioning of edges within a graph or a network. It involves assigning relative positions to edges and determining their visual representation in a graph drawing. The purpose of edge alignment is to enhance the readability and comprehension of the graph by organizing and presenting the edges in a meaningful and structured manner. Edge alignment is crucial in graph visualization, as it influences the perception and understanding of node relationships and connections. By aligning edges in a specific way, certain graph patterns, structures, or properties can be highlighted, making it easier for users to interpret the information conveyed.

One common technique for edge alignment is to use parallel or perpendicular arrangements. In a parallel alignment, edges are drawn parallel, emphasizing their similarity or relationship. This alignment is often used when the edges represent similar or related concepts or have similar properties. For example, consider a graph illustrating a social network where the edges represent friendships. By aligning the edges between friends parallel to each other, the graph can visually convey clusters or communities of friends within the network.

On the other hand, perpendicular alignment involves drawing edges perpendicular to each other. This alignment is typically employed when the edges represent different types of relationships or when there is a hierarchical structure in the graph. For instance, in a graph illustrating a company's organizational structure, the edges between managers and employees can be aligned perpendicularly to the edges between employees and their subordinates, highlighting the hierarchical relationships within the organization.

Edge alignment can also be achieved through various other techniques, such as using curved or straight edges, adjusting the spacing between edges or aligning edges with specific nodes or landmarks in the graph. These techniques can further improve the clarity and understanding of the graph layout.

To illustrate the concept of edge alignment, consider a simple graph with four nodes labeled $A$, $B$, $C$, and $D$ and six edges connecting these nodes in Fig. 3.5. In this example, let us align the edges using parallel and perpendicular arrangements. We can choose to align the edges $\{A, B\}$, $\{A, C\}$, and $\{A, D\}$ parallel to one another to emphasize their similarity, as they all originate from node $A$.[4] The resulting graph with parallel edge alignment would look like Fig. 3.6:

[4] Note that the color red represents the edges that are not aligned with parallel alignment.

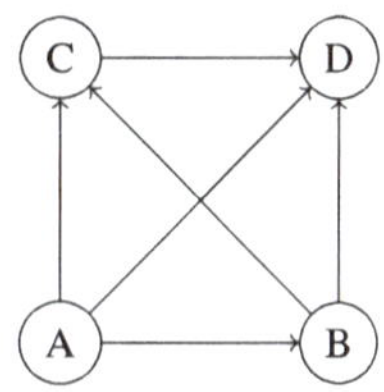


**Fig. 3.5** A simple directed graph with four vertices

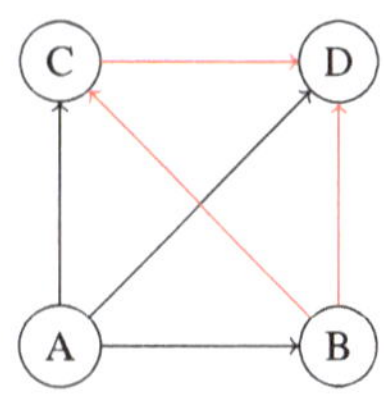


**Fig. 3.6** Parallel alignment of the edges $\{A, B\}$, $\{A, C\}$, and $\{A, D\}$

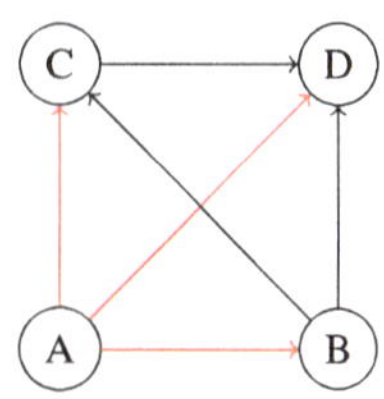


**Fig. 3.7** Perpendicular alignment of the edges $\{A, B\}$, $\{B, C\}$, and $\{C, D\}$

On the other hand, if we choose to align the edges $\{A, B\}$, $\{B, C\}$, and $\{C, D\}$ perpendicularly to the edges $\{A, C\}$, $\{A, D\}$, and $\{B, D\}$, respectively, the resulting graph with perpendicular edge alignment would look like Fig. 3.7:
These examples demonstrate how edge alignment can be used to visually represent different relationships or structures within a graph, aiding in the interpretation and analysis of the underlying data.

In our context, creating a dataset that includes each citation from each paper to understand all the links between entities is crucial. The following example can clarify the steps we have performed.

**Example.**
The paper "The optimal elasticity of taxable income" (SCOPUS-ID: 0036101563) has two authors, Kopczuk (in this example, Author A) and Slemrod (Author B). In their paper, among all the references, they have cited the paper "Tax evasion and optimal commodity taxation" (SCOPUS-ID: 38249007689), which also has two authors, Cremer (Author C) and Gahvari (Author D). Four edges exist between the authors in this example:

1. Author A has cited Author C.
2. Author A has cited Author D.
3. Author B has cited Author C.
4. Author B has cited Author D.

These links are the same for Countries and Affiliations, referring to the authors' details. Analyzing journals' connections, we can observe that both papers have been published on *Journal of Public Economics*. It means that, even if an arrow (i.e., an edge) does not exist on the graph, the degree of the node *Journal of Public Economics* increases by 2. That is, the journal is linked with itself.

By taking advantage of the extensive interconnections among the authors, which also implies connections between countries and affiliations and the links between journals, we aim to construct a comprehensive dataset that captures the relationships between papers and their references. This dataset is formed by 59,264 rows, where each row represents a link between a paper and its corresponding reference, as illustrated in the example provided earlier. To effectively analyze this dataset, we construct networks that focus on isolating the *source* and *destination* nodes of each edge.

It is important to note that the networks we create are directed graphs, meaning that the edges have a specific directionality. This distinction is detailed in Sect. 3.2. By considering the directed nature of the graphs, we can precisely identify the source and destination nodes for each edge, allowing for a more nuanced analysis of the relationships within the dataset. Expanding on the goal of this endeavor, the primary objective is to uncover and explore the intricate connections between authors, countries, affiliations, journals, papers, and references. By leveraging this wealth of information, we can gain insights into the collaboration patterns among the authors, the international reach and impact of research through country and affiliation connections, and the interplay between journals and the papers they publish.

Moreover, by building networks that focus on each edge's source and destination nodes, we can study various aspects of the dataset in a more targeted manner. For instance, we can analyze the citation patterns for specific papers by considering their outgoing edges (source nodes) and incoming edges (destination nodes). This approach allows us to investigate the influence of particular papers on subsequent research or identify pivotal references that have significantly impacted a specific domain.

### 3.3.3 *Visualizing a Network*

Visualizing a network is crucial for gaining insights and understanding complex relationships within the data. By providing a visual representation, the human brain can leverage its innate ability to process visual information more efficiently, leading to improved comprehension, analysis, and interesting and deliverable insights. In particular, visualizing a graph allows us to perceive patterns, clusters, trends, and outliers that might not be immediately evident in raw data. It helps identify the graph's important nodes, influential connections, and structural properties, enabling us to extract meaningful information and make informed decisions.

Moreover, visualizations enhance the communication of graph-based insights to a broader audience who might not be domain experts. By presenting the graph in a visually appealing and intuitive manner, complex concepts and relationships can be conveyed more effectively. Visualizations facilitate storytelling and support data exploration, enabling users to interact with the graph, zoom in on specific areas, and discover hidden patterns or anomalies.

Graph visualization encompasses various methods and techniques to visually represent and display graph structures. These methods play a vital role in analyzing and understanding complex relationships within the data. Some of the most relevant graph visualization methods include force-directed layouts, hierarchical layouts, and matrix-based representations.

The force-directed layout is one of the most widely used approaches conceived in the 60s (Tutte, 1963). Traditionally, this method simulates a physical system where nodes are treated as charged particles and edges as springs with an assumed Hooke's law (Fruchterman and Reingold, 1991). The layout iteratively adjusts the positions of nodes based on attractive forces between connected nodes and repulsive forces between all nodes. The resulting graph layout balances minimizing edge crossings and reducing node overlaps. Mathematically, the force-directed layout can be formulated as an optimization problem, where the objective is to minimize the total energy of the system:

$$E = \sum_{i<j} k \left( \frac{1}{d_{ij}} - \frac{1}{C} \right)^2, \tag{3.8}$$

where $k$ is a constant representing the spring stiffness, $d_{ij}$ is the Euclidean distance between nodes $i$ and $j$, and $C$ is a constant controlling the desired average edge length.

Hierarchical layouts, such as that of Reingold and Tilford (1981), are another important graph visualization method, particularly suited for graphs with inherent hierarchical structures, such as organizational charts or biological taxonomies. These layouts arrange the nodes in a tree-like structure, reflecting the hierarchy. The most common hierarchical layout is the "top-down" or "upward" layout, where nodes are positioned in layers, and edges flow from higher to lower layers. This type of layout aids in visualizing parent–child relationships and hierarchies. Mathematical techniques, such as tree or graph traversal algorithms, can be employed to determine the hierarchical structure and position nodes accordingly.

Matrix-based representations are another effective method for visualizing graphs. This approach represents the graph as an adjacency matrix or an adjacency list, where rows and columns correspond to nodes, and entries indicate the presence or absence of edges. This representation provides a compact and structured view of the graph. Additionally, it enables matrix operations and algorithms for graph analysis, such as computing graph spectra or finding communities. Mathematically, the adjacency matrix $\mathbf{A}$ of an undirected graph with $n$ nodes can be defined as

$$A_{ij} = \begin{cases} 1, & \text{if there is an edge between nodes } i \text{ and } j \\ 0, & \text{otherwise.} \end{cases} \tag{3.9}$$

These are just a few examples of the most relevant graph visualization methods. Other techniques include circular layouts, space-filling layouts, and interactive visualizations. The choice of visualization method depends on the characteristics of the graph and the specific analysis goals. Analysts and researchers can effectively explore, interpret, and communicate complex relationships within graph data by leveraging these methods.

#### 3.3.3.1 The Stress Majorization Algorithm

In the present monograph, to effectively visualize a network, we will use the *stress majorization* algorithm (Gansner et al., 2005). This method is a powerful approach to graph visualization that improves upon traditional force-directed layout methods by minimizing stress, a measure of the dissimilarity between the desired edge lengths and the actual distances in the layout. This method produces visually appealing and informative graph layouts by iteratively adjusting node positions based on stress optimization.

The primary challenge encountered in stress majorization falls within the realm of optimization. The goal is to find an optimal layout that minimizes the stress function, quantifying the discrepancy between the desired and actual distances between pairs of nodes. The stress majorization algorithm formulates the optimization problem as follows:

$$\min_{\mathbf{X}} \sum_{i<j} w_{ij}(d_{ij} - \|\mathbf{x}_i - \mathbf{x}_j\|)^2. \tag{3.10}$$

Here, $\mathbf{X}$ represents the matrix of node positions, where each row $\mathbf{x}_i$ corresponds to the position of node $i$. The term $d_{ij}$ denotes the desired or target distance between nodes $i$ and $j$, and $\|\mathbf{x}_i - \mathbf{x}_j\|$ represents the Euclidean distance between the positions of the nodes. The weights $w_{ij}$ are typically determined by the importance or significance of the edge connecting nodes $i$ and $j$.

To solve this optimization problem, stress majorization utilizes a majorization–minimization approach. At each iteration, the stress function is upper-bounded by a surrogate function that can be minimized analytically. This surrogate function is constructed by linearizing the Euclidean distances in the stress function. By iteratively minimizing the surrogate function, the algorithm updates the node positions until convergence.[5]

Compared to traditional force-directed layout methods, stress majorization offers several advantages. First, it produces layouts that tend to have better edge length preservation. The stress function directly captures the desired distances between nodes, allowing stress majorization to preserve the desired edge lengths in the layout effectively. This feature is handy for graphs with specific distance constraints or when the desired distances convey important information. Second, stress majorization often leads to improved stability and convergence properties. Due to attractive and repulsive force interplay, traditional force-directed methods may suffer from oscillations or slow convergence. In contrast, stress majorization uses an optimization framework that provides a stable and robust approach to finding an optimal layout. Third, stress majorization allows for incorporating additional constraints and objectives into the optimization framework.[6] Finally, stress majorization offers a mathematically sound framework for graph layout optimization. The majorization–minimization approach guarantees convergence to a local minimum of the stress function, providing theoretical guarantees for the quality of the resulting layout. Visualizations are made possible by the R package `ggraph` (Pedersen, 2022), whereas the stress majorization algorithm is implemented in the package `igraph` (Csardi & Nepusz, 2006).

### *3.3.4 Period Split and Comparative Analysis*

To provide additional value to the analysis, we have decomposed the networks studying their evolution across different periods. In particular, each paper has been classified in one of the following periods, according to its publishing date: (1) from 2000 to 2005, (2) from 2006 to 2010, (3) from 2011 to 2015, and (4) from 2016 to the first quarter of 2023.

Decomposing the networks allows us to understand the edges between different actors, the evolution of these links, and the transformation of the network as a whole. As a general trend, largely predictable, networks have been consolidating, fostering knowledge exchange, and drastically reducing cross-country barriers over the years. We will discuss the detailed analysis and results in the following sections. Here, it is essential to mention that the analysis is cumulative (e.g., the second-period analysis includes all the papers from 2000 to 2010).

[5] For an alternative solution that uses additional constraints, we refer the reader to the work of Dwyer et al. (2006).

[6] For example, constraints on node positions, edge weights, or desired distances can be easily incorporated into the stress function, enabling more flexibility in graph visualization. This feature makes stress majorization adaptable to various visualization scenarios and requirements.

## 3.4 Summary

This chapter provides the reader with the fundamental mathematical tools and techniques to analyze networks. After an introduction to graph theory, providing a foundation for understanding the mathematical aspects of networks, it delves into the definition of a graph. It then highlights important properties associated with graphs, which serve as the building blocks of network analysis.

The chapter also covers essential steps such as data cleaning to ensure accurate and reliable network representation. It discusses the alignment of nodes and edges, which is crucial for establishing meaningful connections within the network. The chapter also emphasizes the significance of visualizing networks to gain insights into their structure and relationships. Moreover, techniques like period split and comparative analysis are also presented, enabling researchers to explore variations and patterns within the network.

The knowledge base presented in this chapter is fundamental to performing any empirical analysis. Chapter 4 takes a practical approach by applying these mathematical tools to real-world network data. By bridging the theoretical foundations with empirical investigations, researchers will gain a holistic understanding of network analysis, combining mathematical rigor with actionable insights derived from empirical observations of tax avoidance research networks. The subsequent chapter will demonstrate the power of this mathematical framework in uncovering hidden patterns, identifying key network components, and extracting valuable information from diverse network datasets. Through this seamless transition, readers will embark on a journey that merges theory with practice, empowering them to leverage mathematical methodologies and empirical analysis to gain a deeper understanding of complex networks such as those found in accounting research, particularly in tax avoidance research.

## References

Abbe, E. (2017). Community detection and stochastic block models: recent developments. *Journal of Machine Learning Research, 18*(1), 6446–6531.

Borgatti, S. P., Mehra, A., Brass, D. J., & Labianca, G. (2009). Network analysis in the social sciences. *Science, 323*(5916), 892–895.

Csardi, G., & Nepusz, T. (2006). The igraph software package for complex network research. *InterJournal, Complex Systems, 1695*(5), 1-9.

Dwyer, T., Koren, Y., & Marriott, K. (2006). Stress majorization with orthogonal ordering constraints. In: *Graph Drawing: 13th International Symposium*, GD 2005, Limerick, Ireland, 12–14 Sept. 2005. Revised Papers 13 (pp. 141–152). Springer.

Erdős, P., Rényi, A., et al. (1960). On the evolution of random graphs. *Publication of Mathematics Institute of Hungarian Academy Sciences, 5*(1), 17–60.

Fruchterman, T. M., & Reingold, E. M. (1991). Graph drawing by force-directed placement. *Software: Practice and Experience, 21*(11), 1129–1164.

Gansner, E. R., Koren, Y., & North, S. (2005). Graph drawing by stress majorization. In: *Graph Drawing: 12th International Symposium*, GD 2004, New York, NY, USA, 29 Sept–2 Oct 2004, Revised Selected Papers 12 (pp. 239–250). Springer.

Holland, P. W., Laskey, K. B., & Leinhardt, S. (1983). Stochastic blockmodels: First steps. *Social Networks, 5*(2), 109–137.

Kleinberg, J. M. (1999). Authoritative sources in a hyperlinked environment. *Journal of the ACM (JACM), 46*(5), 604–632.

Kolaczyk, E. D., & Csárdi, G. (2020). Statistical models for network graphs. In: *Statistical Analysis of Network Data with R. Use R!*, Springer, Cham. https://doi.org/10.1007/978-3-030-44129-6_6

Lusher, D., Koskinen, J., & Robins, G. (2013). *Exponential random graph models for social networks: Theory, methods, and applications*. Cambridge University Press.

Muschelli, J. (2019). *rscopus: Scopus database 'API' interface*. R package version 0.6 (p. 6)

Pedersen, T. L. (2022). *ggraph: An Implementation of Grammar of Graphics for Graphs and Networks*. R package version 2.1.0.

Reingold, E. M., & Tilford, J. S. (1981). Tidier drawings of trees. *IEEE Transactions on Software Engineering, 7*(2), 223–228.

Tutte, W. T. (1963). How to draw a graph. *Proceedings of the London Mathematical Society, 3*(1), 743–767.

# Chapter 4
# Networks of Tax Avoidance Research

## 4.1 Introduction

This chapter focuses on the empirical investigation using the network analysis conducted on tax avoidance research. This section discusses the sample selection process, descriptive statistics, and network analysis techniques used to uncover patterns within tax avoidance research. By leveraging the Scopus API and employing various network analysis methods, researchers comprehensively understand the landscape of such an interesting accounting research field.

Section 4.2 begins by describing the sample selection process and the utilization of the Scopus API for data retrieval. Eligibility criteria are established to ensure the relevance and quality of the research included in the analysis. This rigorous selection process forms the foundation for subsequent empirical investigations.

Section 4.3 presents the relevant descriptive statistics. The section examines research productivity by country, institutions, journals in the Chartered Association of Business Schools (hereafter ABS) list, and individual authors. By exploring these statistics, researchers can discern patterns, identify influential contributors, and gain insights into the research landscape.

Section 4.4 delves into network analysis techniques applied to tax avoidance research. This includes the network analysis of countries, affiliations, authors, and journals. Through these analyses, researchers can unveil intricate connections, collaborations, and information flow within the tax avoidance research community.[1] The network analysis provides a powerful lens to identify key players, influential research clusters, and potential research gaps.

Finally, Sect. 4.5 focuses specifically on the European network, presenting a case study that demonstrates the application of data cleaning and alignment techniques.

[1] It is worth highlighting that social networks are important mechanisms for information flow also outside the academic community (Cohen et al., 2008).



A. De Vito, F. Grossetti, *Tax Avoidance Research*, SIDREA Series in Accounting and Business Administration, https://doi.org/10.1007/978-3-031-51765-5_4

The network analysis of European countries, affiliations, authors, and journals within the tax avoidance research domain offers deeper insights into the dynamics and interconnections specific to this region.

## 4.2 Sample Selection and the Scopus API

The interest in tax avoidance has gained momentum over the past decades, as has the research published in academic journals. Considering every paper would lead us to a sample of tens, if not hundreds, of thousands of scientific studies spanning several fields, topics, and journals quality. Hence, it is fundamental to clearly define a series of *eligibility criteria* to narrow the scope of our research to academic papers relevant to the network of accounting scholars.

### *4.2.1 Eligibility Criteria*

Eligibility criteria are a set of rules that researchers use to select participants for their study. In this monograph, a participant will correspond to a scientific paper. These criteria ensure that the sample is representative of the population under investigation. In other words, eligibility criteria ensure that researchers select only participants most likely to provide helpful information to answer the research question. This monograph analyzes the evolution of tax avoidance research over the past two decades and investigates the networks of academics working in such a field.

Eligibility criteria can be divided into two categories: inclusion and exclusion. Inclusion criteria define a set of characteristics for participants to be included in the study. In accounting research, an example of an inclusion criterion could be the firm's size. A researcher studying the relationship between audit quality and earnings management may be interested in selecting only listed companies with market capitalization equal to or higher than 100 million USD. This criterion will include only larger companies whose financial statements are likely mandated to be audited. Other examples of inclusion criteria in accounting research could be the industry (e.g., financial services, manufacturing, retail, etc.), the geographic location (e.g., Europe versus the United States), or the section of the financial statement being analyzed (e.g., income statement, balance sheet, etc.).

In this work, we define two main inclusion criteria to correctly identify a suitable sample that would enable us to answer our research question. First, we include only studies published from 2000 to the first quarter of 2023. This window allows us to dynamically study the field's evolution over the past two decades. Second, we follow prior literature (Hanlon & Heitzman, 2010; Wilde & Wilson, 2018; Jacob, 2022) and use the keywords reported in Table 4.1.

**Table 4.1** List of keywords used to identify tax avoidance-related papers in Scopus

| ID | Keyword |
|---|---|
| 1 | Tax avoidance/corporate tax avoidance |
| 2 | Tax aggressiveness |
| 3 | Tax planning/corporate tax planning |
| 4 | Tax shelter/tax sheltering |
| 5 | Profit shifting |
| 6 | Income shifting |
| 7 | Transfer pricing |
| 8 | Tax haven/tax havens |
| 9 | Debt shifting |
| 10 | Effective tax planning |

As it can be inferred from the list above, the keywords provide comprehensive coverage of the research on tax avoidance. The initial sample selection yields 4562 papers, 6587 authors, and 2787 institutions, which were obtained through the official Elsevier Scopus API.[2]

### 4.2.2 *The Scopus API*

The Elsevier Scopus Application Programming Interface (or API) is a software system that allows two applications to interact without user intervention. It can be publicly accessed and executed remotely upon registration. We chose Elsevier Scopus API for our analysis because it provides access to abstract and citation data from all scholarly journals formally indexed by Scopus.[3] To use an API, the researcher needs an API key, a unique identifier that processes requests associated with a running project.

As a front end, we use the R package `rscopus` (Muschelli, 2019), which provides an easy and direct interface to query the database in an SQL-like fashion. Figure 4.1 reports the query and the code we used to extract the initial sample. Each keyword was chained using an `OR` condition to ensure maximum flexibility when retrieving papers. For example, a paper containing the keyword "tax shelter" will have the same likelihood of being selected as a paper containing two keywords "tax shelter" and "profit shifting."

The query in Fig. 4.1 yields a large sample of 4562 papers, 6587 authors, and 2787 institutions, as mentioned above. Using large samples can enhance the

[2] The interested reader can find more about the Scopus API at https://dev.elsevier.com.

[3] Researchers in academic, public sector, and not-for-profit institutions can use Scopus APIs for non-commercial use at no additional cost. Some examples of Elsevier Scopus API include Scopus Search API, which covers 50 million abstracts of over 20,500 peer-reviewed titles from more than 5000 publishers, and embedding Scopus Cited-by counts on your site.

```
tax_avoidance_query <- scopus_search(
  "TITLE-ABS-KEY(\"tax avoidance\")
                                   OR TITLE-ABS-KEY(\"corporate tax avoidance\")
                                   OR TITLE-ABS-KEY(\"tax aggressiveness\")
                                   OR TITLE-ABS-KEY(\"tax planning\")
                                   OR TITLE-ABS-KEY(\"corporate tax planning\")
                                   OR TITLE-ABS-KEY(\"tax shelter\")
                                   OR TITLE-ABS-KEY(\"tax sheltering\")
                                   OR TITLE-ABS-KEY(\"profit shifting\")
                                   OR TITLE-ABS-KEY(\"income shifting\")
                                   OR TITLE-ABS-KEY(\"transfer pricing\")
                                   OR TITLE-ABS-KEY(\"tax haven\")
                                   OR TITLE-ABS-KEY(\"tax havens\")
                                   OR TITLE-ABS-KEY(\"debt shifting\")
                                   OR TITLE-ABS-KEY(\"effective tax planning\")
                                   AND PUBYEAR > 1999",
  view = "COMPLETE",
  headers = insttoken
)
```

**Fig. 4.1** The initial query to Scopus Search API executed with the function `scopus_search()` from the package `rscopus`

statistical power of empirical analysis by providing more precise estimates and increasing the likelihood of detecting true effects or relationships. However, it is essential to recognize that working with such large settings may introduce particular challenges. These challenges include increased computational and logistical demands, necessitating adequate infrastructure and resources. Additionally, attention must be given to ensuring the quality and representativeness of the data, as well as addressing potential biases. Furthermore, the data analysis and interpretation complexity can escalate when dealing with large samples, requiring advanced techniques to extract meaningful insights. Among the presented issues, we highlight three that are of great importance for our scope:

**Invalidity:** The data may not answer the research question or may answer a different question from the researcher had previously intended.

**Irrelevance:** The data may not be useful for the intended purpose or audience.

**Inaccuracy:** The data may have errors or inconsistencies that affect the quality and reliability of the analysis.[4]

These issues can severely impair the empirical analysis and the results. Hence, to overcome them, we carefully define and impose additional eligibility criteria for the data selection and cross-check the data before performing any analysis. First, we define all relevant publication outlets considered in our network analysis. Given our scope, we limit the search to studies published in top accounting, finance,

[4] This last point should not be a concern as we are directly interacting with a high-quality database provided by Scopus.

and economics journals.[5] To do so, we refer to the internationally recognized Academic Journal Guide 2021 compiled by the Chartered Association of Business Schools (i.e., the ABS list), a journal ranking widely used for tenure promotion (Bajo et al., 2020).[6] From this list, we select all journals with a 3, 4, and 4* star classification in the ABS list because the research appearing in these journals is likely of high or very high quality.[7] To further validate this criterion, we cross-check our list with the A-level Journal List of area 13 compiled by the Italian's *Agenzia Nazionale di Valutazione del Sistema Universitario e della Ricerca* (ANVUR) (hereafter ANVUR list).[8] Furthermore, we exclude all the observations for which any of the following API fields are missing: `auth-id`, `eid`, `dc:title`, `prism:publicationName`, `prism:coverDate`, and `afid`. Applying these stricter criteria reduces our initial sample to top 612 publications.

However, we acknowledge that our strict criteria may entail some limitations. First, our approach excludes unpublished working papers, which may seemingly be important for moving the literature forward. Nevertheless, we conservatively include only published studies in our sample because of the uncertainty around the publication process. Second, we understand that other studies may have been published before 2000. Reassuringly, though, in untabulated robustness tests, we apply similar search criteria and are able to add 31 papers to our sample. We repeat all the analyses and find unchanged results (available upon request).

## 4.3 Descriptive Statistics

The final sample of 612 publications includes 1049 unique authors from 2000 to the first quarter of 2023 (see Table A.1 for the full list of the sample publications). The average (median) number of authors per paper is 2.52 (3). Moreover, each paper receives an average (median) of 34.27 (10) citations. In Table 4.2, we show the differences in the number of authors per paper and the number of citations per paper, conditional on the ABS classification as defined in the eligibility criteria.

[5] We acknowledge that tax avoidance research could also appear in top journals of other disciplines. However, we believe this should not be a major concern for our analyses because we purposely narrow down the analysis to key topics unlikely to be of interest to non-accounting scholars.

[6] For additional information on the ABS Journal Guide, we refer the reader to https://charteredabs.org/academic-journal-guide-2021/.

[7] For example, the ABS list states that all journals falling under category 3 are "*very selective in what they publish.* [And the] *papers are heavily refereed.*" Similarly, the journals falling under category 4 "*are top journals in their field* [and] *typically have high submission and low acceptance rates.*" Finally, the journals under category 4* "*are recognized worldwide as exemplars of excellence.*" We also note that the journals in our list belong to either the first or the second quartile of the Scimago journal rank, which further signals that our coverage only comprises top accounting, finance, and economics journals that likely drive the field.

[8] Specifically, we use the ANVUR list 2023.

**Table 4.2** Average and median number of authors per paper for top ABS rankings

| | Number of authors | | Number of citations | |
|---|---|---|---|---|
| ABS classification | Mean | Median | Mean | Median |
| 3 | 2.42 | 2 | 24.02 | 8 |
| 4 | 2.68 | 3 | 32.22 | 12 |
| 4* | 2.83 | 3 | 91.27 | 19 |

**Table 4.3** Top 20 countries for research productivity

| Country | Publications |
|---|---|
| United States | 737 |
| Germany | 111 |
| United Kingdom | 99 |
| China | 77 |
| Canada | 74 |
| Australia | 73 |
| Hong Kong | 58 |
| Netherlands | 40 |
| France | 28 |
| Spain | 23 |
| Singapore | 22 |
| Taiwan | 21 |
| Denmark | 18 |
| Sweden | 15 |
| Finland | 14 |
| Norway | 12 |
| Austria | 11 |
| Italy | 11 |
| Ireland | 10 |
| Switzerland | 9 |

Interestingly, and somewhat expected, we note that publishing in highly ranked journals results in more citations.

### *4.3.1 Research Productivity by Country*

Next, we analyze the research productivity along several dimensions. We start by looking at the number of publications per country. Table 4.3 shows the top 20 countries based on the number of publications. The United States is the first country with 737 publications, followed by Germany with 111 and the United Kingdom with 99. Italy is tied to Austria in the 18th place, with each country having 11 publications. However, it is worth noting that 12 out of the top 20 are European countries. These include Germany, the United Kingdom, the Netherlands, France,

Spain, Denmark, Sweden, Finland, Norway, Austria, and Italy. This suggests that tax avoidance research extends beyond the confines of the United States, encompassing a broader international scope.

Table 4.3 underscores the importance of international collaboration and knowledge exchange in driving scientific progress, as researchers worldwide can learn from and build upon each other's work.

### *4.3.2 Research Productivity by Institutions*

We now turn our attention to the most productive research institutions. To have a reliable overview of the institutions, we manually aggregate business schools to their parent universities whenever multiple entities with different names belong to the same research institution. For example, Bocconi University and SDA Bocconi School of Management will be under the "Bocconi System" umbrella. Hence, the term *System* denotes a collection of entities within the same institution considered at large. Table 4.4 reports some descriptive statistics. In line with the previous analysis, US institutions dominate the list, even though we find one European school in the ninth position (i.e., Ludwig-Maximilians-Universität München). Consistent with Table 4.3, the results suggest that US Institutions are highly productive in the tax avoidance field.

This is an interesting initial overview as it highlights the top institutions where tax avoidance research is conducted and published in leading journals. Unsurprisingly, we observe the United States as the front-runner, as the whole field went from a niche area to mainstream research after the seminal work of Scholes et al. (2014) and the influential reviews of Shackelford and Shevlin (2001) and Hanlon and Heitzman (2010).

**Table 4.4** Research productivity of the top ten institutions in our sample. The term *System* reflects the authors' aggregation to account for several entities belonging to the same parent university

| Affiliation | Publications |
|---|---|
| The University of California System | 37 |
| State University System of Florida | 26 |
| The University of Texas System | 26 |
| Texas A and M University System | 24 |
| The Regents of the University of Michigan | 24 |
| University System of Georgia | 22 |
| The University of Wisconsin System | 21 |
| The University of Illinois System | 20 |
| Ludwig-Maximilians-Universität München | 19 |
| Oregon University System | 18 |

**Table 4.5** Number of papers published in the top 15 ABS journals classified as 3

| ABS journals—level 3 | Publications |
|---|---|
| Journal of Public Economics | 67 |
| Journal of the American Taxation Association | 60 |
| Journal of Business Ethics | 36 |
| Journal of International Accounting, Auditing and Taxation | 26 |
| European Accounting Review | 23 |
| Accounting Forum | 18 |
| Accounting and Business Research | 18 |
| Critical Perspectives on Accounting | 15 |
| Journal of Accounting and Public Policy | 15 |
| Journal of Accounting, Auditing, and Finance | 15 |
| Review of Quantitative Finance and Accounting | 15 |
| Journal of Business Finance and Accounting | 14 |
| British Tax Review | 12 |
| American Economic Journal: Economic Policy | 10 |
| Scandinavian Journal of Economics | 7 |

### 4.3.3 Research Productivity by Journal in the ABS List

We now explore the research output composition based on the different ABS journal classifications. The findings are reported in Table 4.5 specifically for ABS journals ranked as 3. As all the publications in the sample focus on tax avoidance research, this table offers valuable insights into the publishing trends within this domain. Notably, the significant number of publications in the *Journal of Public Economics* and the *Journal of the American Taxation Association* suggests that these specialized journals hold substantial influence and enjoy wide readership in tax avoidance research. Therefore, researchers interested in contributing to tax-related topics may consider submitting their work to these journals. Additionally, it is worth highlighting that the *Journal of Business Ethics*, which centers around the ethical implications of tax avoidance, ranks third in publications. This finding implies that investigating the ethical dimension of tax avoidance holds significance for policymakers and academics alike (Dowling, 2014). In conclusion, the table provides valuable information to researchers exploring tax avoidance research. By comprehending the publishing trends in this domain, researchers can gain insights into the most influential journals and the topics currently receiving the most attention.

We now replicate the analysis for ABS 4 and 4* journals and report the results in Tables 4.6 and 4.7, respectively. Unlike Tables 4.5, 4.6 encompasses journals with a broader range of topics, including accounting, economics, and finance. *Contemporary Accounting Research* is the leading journal in this table, which boasts 55 publications. It holds a prestigious position within the accounting field and is recognized for publishing research that carries significant implications for real-world

**Table 4.6** Number of papers published in the top ABS 4 journals

| ABS journals—level 4 | Publications |
|---|---|
| Contemporary Accounting Research | 55 |
| Review of Accounting Studies | 18 |
| Journal of International Economics | 7 |
| Production and Operations Management | 5 |
| Journal of Financial and Quantitative Analysis | 4 |
| Review of Economics and Statistics | 3 |
| International Economic Review | 2 |
| Journal of Management Studies | 2 |
| Journal of the European Economic Association | 2 |
| American Economic Journal: Macroeconomics | 1 |
| Economic Journal | 1 |
| Human Relations | 1 |

**Table 4.7** Number of papers published in the top ABS 4* journals

| ABS journals—level 4* | Publications |
|---|---|
| Journal of Accounting Research | 17 |
| Accounting Review | 15 |
| Journal of Financial Economics | 13 |
| Journal of International Business Studies | 12 |
| Management Science | 10 |
| Strategic Management Journal | 4 |
| Quarterly Journal of Economics | 3 |
| American Economic Review | 2 |
| Review of Economic Studies | 2 |
| Review of Financial Studies | 2 |
| Journal of Finance | 1 |

applications. The *Review of Accounting Studies* follows closely in the second place with 18 publications, also renowned for its contribution to accounting scholarship. The third journal, the *Journal of International Economics*, is an economics-focused journal that caters to international tax and finance scholars. Contributions to this journal focus more on *how* firms avoid taxes (e.g., through income shifting) rather than *whether* firms engage in tax avoidance behavior. The remaining journals exhibit fewer publications, ranging from 5 to 1. Overall, this table indicates the continued prominence of accounting research within the ABS journals classified as 4, with two out of the top three journals belonging to the accounting discipline. However, including journals from diverse fields like economics and finance again underscores the interdisciplinary nature of tax avoidance research (Hanlon & Heitzman, 2010).

Table 4.7 presents the publication count in the top ABS 4* journals, denoting exceptional quality. Notably, only a handful of journals on this list specifically concentrate on tax research. Instead, these journals encompass various subjects within the broader business administration field, including accounting, finance, management, and corporate strategy. The *Journal of Accounting Research* and *The*

*Accounting Review* are the sole journals on the list dedicated to pure accounting research. In contrast, the *Journal of Financial Economics*, the *Journal of Finance*, and the *Review of Financial Studies* hold prominence in finance. The *Journal of International Business Studies* explores international business research, while *Management Science* and the *Strategic Management Journal* delve into management and corporate strategy studies. Notably, the *Quarterly Journal of Economics*, the *American Economic Review*, and the *Review of Economic Studies*, all esteemed economics journals, cover a broad spectrum of topics.

Although the number of journals focusing explicitly on tax research is limited within this list, it is worth highlighting that tax-related studies can still find a place for publication. This suggests that tax research extends beyond a limited selection of journals and intersects with broader research streams. Furthermore, the scarcity of tax-specific journals within this list indicates that tax research transcends disciplinary boundaries and encompasses various disciplines. Taxation, as a research field, spans tax-specific journals and permeates multiple academic domains. This underscores the broad nature of tax research, with its connections to accounting, finance, economics, and law, contributing to a multifaceted exploration of the topic (Hanlon & Heitzman, 2010).

### 4.3.4 Research Productivity by Author

We now shift our focus to examining the production function of authors in the field. Following the approach used for journals, we present a comprehensive analysis of the top 20 authors within each ABS category, as illustrated in Table 4.8. This table encompasses the authors published in journals with ABS journals classified as 3, 4, and 4*. Upon examining the list of authors, it becomes evident that established and early career academics seemingly contribute to tax avoidance research. With their extensive publication histories, renowned scholars such as Slemrod J., Wilson R., and Omer T. feature alongside rising talents like Laplante S. K. and Gaertner F. B., who have made noteworthy contributions. Exploring affiliations, we observe a strong presence of authors from the United States. However, it is worth noting the inclusion of authors from other countries, such as Germany (Wamser G. and Riedel N.), highlighting the global nature of tax avoidance research and the collaborations that span international borders.

Interestingly, many authors listed in the ABS journals classified as 4 and 4* hold esteemed positions as leading scholars in the field of tax research. For instance, Shevlin T., Maydew E. L., and Dyreng S. are widely recognized for their influential work on tax avoidance and tax policy. Similarly, Wilson R., Thornock J. R., and Hoopes J. are respected scholars in tax compliance and enforcement. Such a characteristic is a typical trait also occurring in other fields of economics where a relatively small number of outstanding scholars co-author with many other academics (Goyal et al., 2006).

**Table 4.8** Top 20 authors in each ABS journal category

| ABS—level 3 | | ABS—level 4 | | ABS—level 4* | |
|---|---|---|---|---|---|
| Author | Publications | Author | Publications | Author | Publications |
| Johannesen N. | 10 | Shevlin T. | 6 | Dyreng S. | 5 |
| Omer T. | 8 | Klassen K. | 5 | Maydew E.L. | 4 |
| Richardson G. | 7 | Laplante S.K. | 4 | Shevlin T. | 4 |
| Zucman G. | 6 | Stomberg B. | 4 | Wilson R. | 3 |
| Slemrod J. | 6 | Gaertner F.B. | 4 | Thornock J.R. | 3 |
| Wilson R. | 5 | McGuire S. | 3 | Wilde J. | 3 |
| Taylor G. | 5 | Watson L. | 3 | Wu Q. | 2 |
| Jacob M. | 5 | Lynch D. | 3 | Hoopes J. | 2 |
| Stekelberg J. | 5 | Lewellen C. | 3 | Lisowsky P. | 2 |
| Khurana I.K. | 5 | Omer T. | 3 | Lindsey B. | 2 |
| Borkowski S.C. | 5 | Maydew E.L. | 3 | Zhang L. | 2 |
| Sikka P. | 5 | Ayers B.C. | 3 | Zhang H. | 2 |
| Lobo G.J. | 4 | Xiao W. | 2 | De Simone L. | 2 |
| Riedel N. | 4 | Wilson R. | 2 | Chen T. | 2 |
| Lanis R. | 4 | Wu Q. | 2 | Lester R. | 2 |
| Argilés-Bosch J.M. | 4 | Lisowsky P. | 2 | Hoi C. | 2 |
| Wamser G. | 4 | Jacob M. | 2 | Reichelstein S. | 2 |
| Doyle E.M. | 4 | Huston G. | 2 | Hasan I. | 2 |
| Wu Q. | 4 | Donohoe M. | 2 | Hanlon M. | 2 |
| McGuire S. | 4 | Blaylock B. | 2 | Mills L. | 2 |

The table provides a valuable snapshot of the most prolific authors in tax avoidance research based on their publications in ABS-ranked journals. While factors such as seniority and affiliation hold importance, the table underscores that the quality and impact of an author's research play a pivotal role in determining their position in the rankings. This highlights the significance of research excellence in the field and reaffirms the recognition of these authors for their contributions.

## 4.4 Networks of Academic Collaboration

After having provided a comprehensive overview of graphs, the measures used to define links and edges, and the eligibility criteria in Chap. 3, we now delve into the empirical results obtained with the network analysis approach (Kong et al., 2019). We first examine the links between the countries from which the authors' affiliations are to gain insights into the connections within the tax avoidance literature. This analysis helps us identify the geographical locations that serve as crucial research hubs in the field, referring to them as the "main" hubs without necessarily implying their relative productivity, as illustrated in Sect. 4.3.

Moreover, leveraging information on the affiliations' locations of the authors in our dataset, we isolate those with European affiliations to conduct the same analyses exclusively within the European subgroup (see Sect. 4.5). This approach allows us to gain a focused understanding of the dynamics specific to the European context, thus further extending and elaborating on critical insights that might be particularly relevant to local scholars and researchers. Furthermore, this focused analysis of the European context can provide valuable guidance and serve as a helpful resource for young accounting scholars (including Ph.D. candidates, Postdoctoral researchers, and early stage Assistant Professors) seeking to identify prominent tax research hubs within Europe, aiding them in selecting suitable institutions for their academic pursuits. Continuing our investigation, we explore the centrality of different affiliations and the significance of individual authors within this stream of literature. By examining the affiliations' centrality, we gain valuable insights into the influential nodes within the network. Lastly, we comprehensively analyze the existing links and their evolution over time among various journals. This examination provides valuable insights into the types of journals, their rankings, and the topics they cover. Hence, it allows us to understand the landscape of tax avoidance research publications in terms of journal characteristics and trends.

Our objective is to facilitate a more profound comprehension of the interconnections, geographical distributions, centrality of affiliations, and evolution of journals in the tax avoidance literature. By uncovering these insights, we contribute to fostering a comprehensive and nuanced understanding of the research landscape in this field, benefiting researchers, policymakers, and stakeholders alike.

### *4.4.1 Network Analysis of Countries*

Figure 4.2 illustrates the country network as of the first quarter of 2023, consisting of 69 interconnected nodes representing different countries. These nodes are linked by 770 unique edges, indicating their connections. The graph shows that the network exhibits a high density, with a notable concentration around the central node, symbolizing the *United States*. The strongest edges, denoted by a darker shade of blue, are predominantly found between the central node and countries such as *United Kingdom*, *Canada*, *Hong Kong*, and *Australia*. Table 4.9 provides further details, highlighting that the *United States* possesses a total degree centrality of 89, indicating the existence of 89 connections between this node and other countries.[9]

This network pattern aligns with expectations. As we delve into the affiliation network analysis, we will observe a strong presence of tax avoidance literature originating from the United States. Over time, this literature has expanded to encompass European countries and Eastern affiliations through various studies and comparative analyses of different tax systems (Fox et al., 2022).

[9] It is important to note that the network is directed, meaning that the link between the United States and another country, as well as its reciprocal, is counted twice.

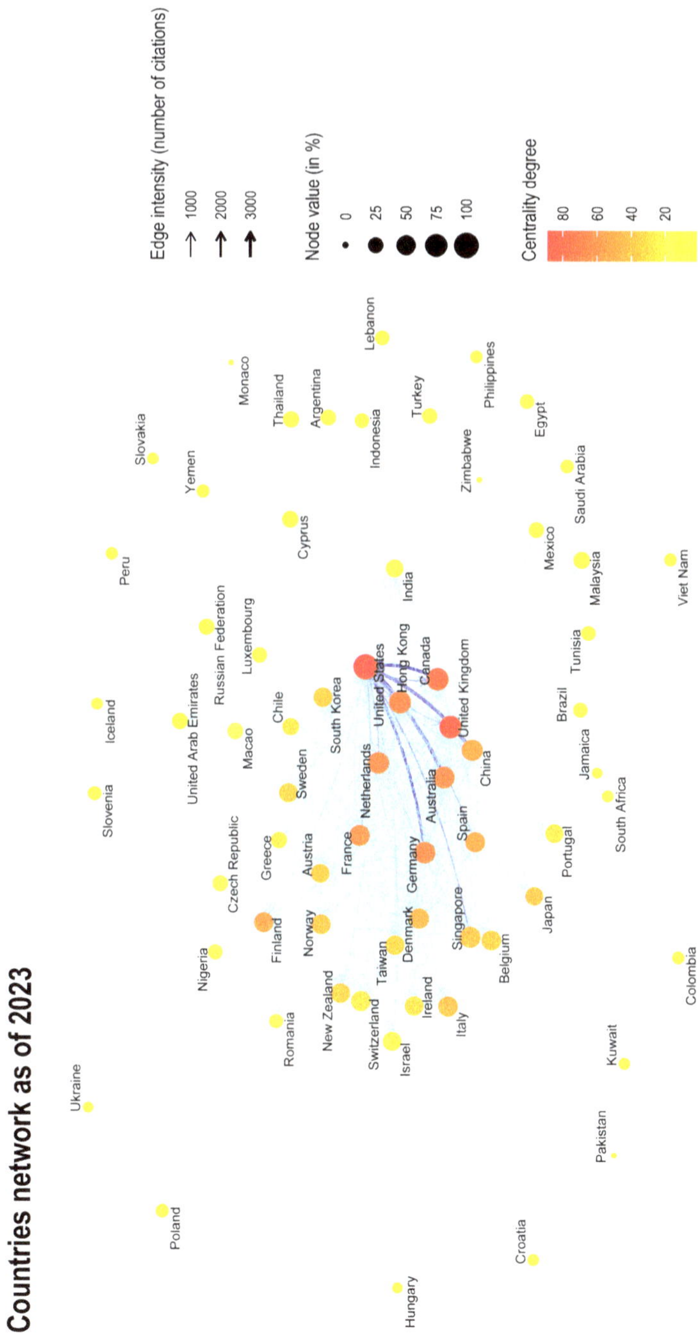


**Fig. 4.2** Graph of the countries as of 2023

**Table 4.9** Descriptive statistics for the network of countries as of 2023

| Node | Country | Degree | Node | Country | Degree | Node | Country | Degree |
|---|---|---|---|---|---|---|---|---|
| *1* | United States | 89 | *24* | Chile | 27 | *46* | Turkey | 5 |
| *2* | United Kingdom | 85 | *41* | Switzerland | 27 | *57* | Indonesia | 5 |
| *4* | Canada | 75 | *25* | Portugal | 26 | *45* | Argentina | 4 |
| *3* | Hong Kong | 65 | *23* | Greece | 25 | *47* | Egypt | 3 |
| *6* | Australia | 63 | *11* | Israel | 24 | *52* | Slovenia | 3 |
| *18* | Germany | 63 | *35* | Romania | 21 | *55* | Lebanon | 3 |
| *9* | Netherlands | 62 | *32* | India | 20 | *63* | Yemen | 3 |
| *20* | France | 60 | *36* | Macao | 17 | *69* | Viet Nam | 3 |
| *26* | Finland | 53 | *28* | Czech Republic | 15 | *48* | Poland | 2 |
| *8* | China | 50 | *14* | Nigeria | 14 | *53* | Kuwait | 2 |
| *17* | Spain | 50 | *33* | South Africa | 14 | *54* | Colombia | 2 |
| *19* | Denmark | 44 | *40* | Luxembourg | 13 | *60* | Philippines | 2 |
| *27* | Italy | 43 | *34* | Brazil | 11 | *66* | Iceland | 2 |
| *10* | Singapore | 40 | *38* | Jamaica | 11 | *68* | Ukraine | 2 |
| *5* | Norway | 39 | *21* | Tunisia | 10 | *42* | Monaco | 1 |
| *15* | Japan | 39 | *37* | United Arab Emirates | 10 | *50* | Hungary | 1 |
| *39* | Belgium | 36 | *49* | Russian Federation | 8 | *56* | Estonia | 1 |
| *7* | South Korea | 35 | *22* | Saudi Arabia | 7 | *58* | Peru | 1 |
| *16* | Austria | 35 | *30* | Zimbabwe | 7 | *61* | Oman | 1 |
| *31* | New Zealand | 35 | *44* | Malaysia | 7 | *62* | Libya | 1 |
| *12* | Sweden | 34 | *59* | Cyprus | 7 | *64* | Pakistan | 1 |
| *13* | Taiwan | 33 | *51* | Mexico | 6 | *65* | Croatia | 1 |
| *29* | Ireland | 30 | *43* | Thailand | 5 | *67* | Slovakia | 1 |

Another meaningful insight, evident from the outer ring of the network plot and detailed in Table 4.9, is the involvement of smaller countries in tax avoidance literature. Countries such as *Monaco*, *Peru*, *Oman*, and *Libya* have gained attention and contributed to the literature by offering unique perspectives. These countries enrich the discourse and leverage existing literature to develop new theories, models, and conclusions.

#### 4.4.1.1 Time Evolution of Countries Network

In studying complex systems, understanding how they evolve is crucial for gaining insights into their behavior and dynamics. This is particularly true in the network analysis, where examining the temporal aspect of networks can provide valuable information about the changing relationships, connectivity, and influence patterns. By analyzing networks across different periods, researchers can uncover trends, detect structural changes, and identify critical events or transitions that shape the system's evolution. We divide our time window into 2005, 2010, and 2015.

In the initial period of our analysis, encompassing papers published from January 1, 2000 to December 31, 2005, we examine 38 research papers. Figure 4.3 reveals an apparent network size and density contrast during this time frame. This observation is further supported by the summary provided in Table 4.10, indicating that only 20 countries had authors who published papers on tax avoidance, adhering to the eligibility criteria outlined in Sect. 4.2.1. Moreover, the connections among these countries were relatively fewer and less pronounced, with a count of only 45

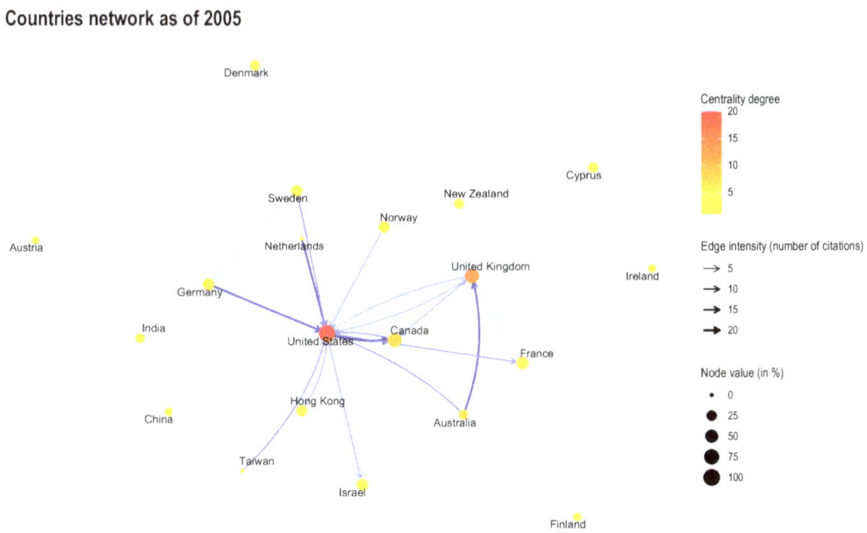


Fig. 4.3 Countries network analysis as of 2005

**Table 4.10** Countries network analysis as of 2005

| Node | Country | Degree |
|---|---|---|
| *1* | United States | 20 |
| 2 | United Kingdom | 13 |
| *10* | Canada | 8 |
| *5* | Australia | 7 |
| *4* | Norway | 6 |
| *8* | Germany | 6 |
| *3* | New Zealand | 5 |
| *7* | Sweden | 5 |
| *11* | Hong Kong | 5 |
| *6* | Netherlands | 3 |
| *9* | Taiwan | 2 |
| *12* | France | 2 |
| *13* | Israel | 1 |
| *14* | China | 1 |
| *15* | Denmark | 1 |
| *16* | Austria | 1 |
| *17* | Finland | 1 |
| *18* | Ireland | 1 |
| *19* | India | 1 |
| *20* | Cyprus | 1 |

blue lines observed. The limited frequency and intensity of connections between countries during this period can be attributed to several factors. First, the field of tax avoidance research could have been relatively less developed, with fewer scholarly contributions available compared to later years. This scarcity of published papers could have resulted in fewer opportunities for collaboration and knowledge exchange among researchers from different countries. Second, the academic community focusing on tax avoidance could have been less interconnected and cohesive during that time, with fewer established networks and platforms for collaboration. Third, the lower frequency of connections could also reflect the limited availability of data and resources, making it challenging for researchers to conduct comprehensive cross-country analyses. Overall, these factors collectively contribute to the relatively sparse and less intense network observed in the earlier period of our analysis.

The centrality analysis reveals that the *United States* remains the most central node in the network, maintaining its prominent position with a centrality degree of 20. Similarly, other countries, such as the *United Kingdom*, *Canada*, and *Australia*, continue to maintain strong connections with the United States. These countries share commonalities, including the same language and a common-law legal system. Notably, all four countries have a profit-based corporate tax system, allowing businesses to deduct expenses from their taxable income, including previous tax losses. Additionally, all four countries are OECD members and have put effort into curbing tax avoidance in recent years. Looking at Table 4.3, we can furthermore

Countries network as of 2010

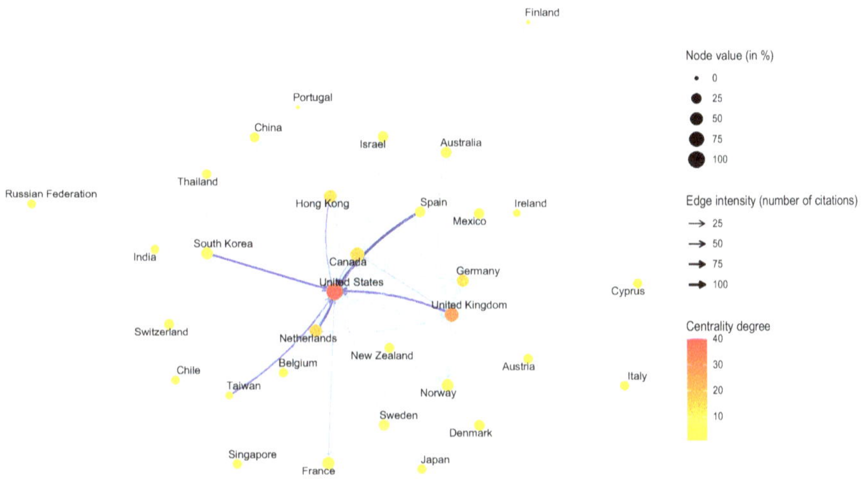


**Fig. 4.4** Countries network analysis as of 2010

discover two other interesting insights: (1) emerging economies like *China* and *India* have been contributing to the literature since the very beginning, and (2) Europe has started to assume a critical role in the spread of tax avoidance literature in the continent. This will again be confirmed by the affiliations based in these countries.

To capture the evolving landscape of tax avoidance research, we now focus on papers published from 2006 to 2010, providing a snapshot of the literature's evolution over an additional 5-year period. This expansion includes 95 papers authored by individuals from 32 countries, leading to 114 connections in the network, more than 2.5 times the number observed in the previous 5-year period.

Examining Fig. 4.4, we once again observe the central role of the *United States* in the network. However, an additional noteworthy finding emerges regarding the overall network structure. As corroborated by Table 4.11, the *United Kingdom* appears as the second most central node in the network, serving as a reference point for numerous European countries such as *Germany*, *Netherlands*, *Austria*, and *Italy*. Notably, *Spain* stands as an exception due to its connections with the *United States*. Collectively, the network indicates a growing closeness among countries over time, as confirmed by the subsequent snapshot in 2015.

The corpus has witnessed a significant growth over the subsequent 5-year period (2011–2015), with 114 new papers published. This expansion results in 209 documents from 40 countries, marking a 25% increase compared to the previous segment. The heightened level of scholarly output is accompanied by a substantial rise in edges, reaching 233, a remarkable 61.8% increase. These findings reaffirm the trend of countries converging toward a more interconnected tax avoidance literature, as reported in Table 4.12.

**Table 4.11** Countries network analysis as of 2010

| Node | Country | Degree | Node | Country | Degree |
|---|---|---|---|---|---|
| *1* | United States | 40 | *27* | Mexico | 5 |
| *3* | United Kingdom | 27 | *14* | China | 4 |
| *8* | Netherlands | 15 | *18* | France | 4 |
| *11* | Germany | 14 | *25* | Denmark | 4 |
| *12* | Canada | 14 | *20* | Austria | 3 |
| *6* | Hong Kong | 13 | *24* | Switzerland | 3 |
| *10* | Australia | 10 | *31* | Thailand | 3 |
| *4* | Spain | 8 | *19* | Japan | 2 |
| *5* | Sweden | 8 | *22* | Chile | 2 |
| *9* | South Korea | 7 | *21* | Italy | 1 |
| *16* | Norway | 7 | *23* | Singapore | 1 |
| *2* | Taiwan | 6 | *26* | Finland | 1 |
| *13* | Belgium | 6 | *28* | India | 1 |
| *15* | New Zealand | 6 | *29* | Cyprus | 1 |
| *7* | Ireland | 5 | *30* | Portugal | 1 |
| *17* | Israel | 5 | *32* | Russian Federation | 1 |

**Table 4.12** Countries network analysis as of 2015

| Node | Country | Degree | Node | Country | Degree |
|---|---|---|---|---|---|
| *1* | United States | 56 | *18* | Ireland | 9 |
| *6* | United Kingdom | 41 | *23* | South Korea | 9 |
| *21* | Germany | 28 | *12* | Singapore | 8 |
| *2* | Canada | 27 | *13* | Italy | 8 |
| *14* | Netherlands | 26 | *27* | Norway | 8 |
| *8* | Hong Kong | 22 | *28* | Switzerland | 8 |
| *10* | Australia | 22 | *24* | Portugal | 7 |
| *9* | Spain | 17 | *33* | Mexico | 5 |
| *15* | Finland | 15 | *7* | Nigeria | 4 |
| *16* | France | 15 | *30* | Malaysia | 4 |
| *19* | Austria | 14 | *39* | Thailand | 3 |
| *3* | Israel | 13 | *29* | United Arab Emirates | 2 |
| *26* | Belgium | 13 | *31* | Argentina | 1 |
| *11* | Denmark | 12 | *32* | Chile | 1 |
| *4* | Taiwan | 11 | *34* | Indonesia | 1 |
| *17* | Sweden | 11 | *35* | Cyprus | 1 |
| *25* | New Zealand | 11 | *36* | Yemen | 1 |
| *20* | China | 10 | *37* | Pakistan | 1 |
| *22* | India | 10 | *38* | Croatia | 1 |
| *5* | Japan | 9 | *40* | Russian Federation | 1 |

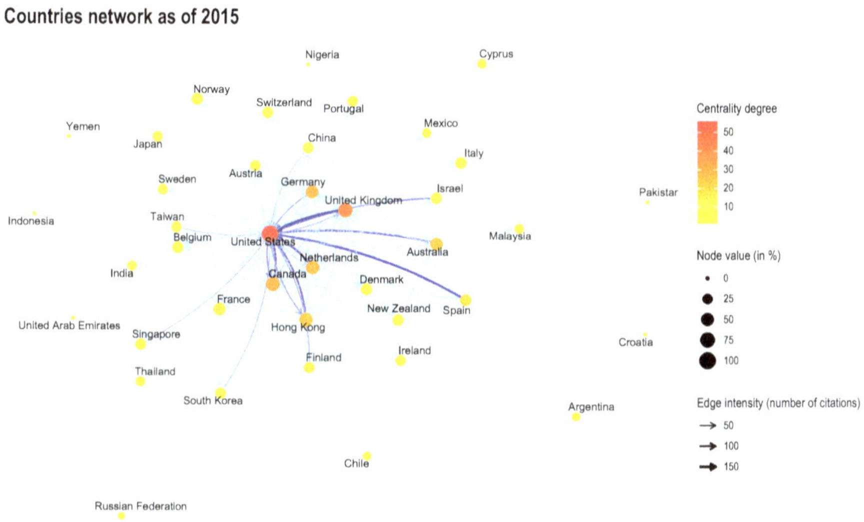


**Fig. 4.5** Countries network analysis as of 2015

Examining the network depicted in Fig. 4.5, we observe a notable shift in the shape of the network, characterized by an increase in interactions among not only the most central nodes but also various connected nodes. One prominent feature is the conspicuous triangle formed by the *United States*, the *United Kingdom*, and the *Netherlands*. The connections between the United States, the United Kingdom, and *Germany* are also noteworthy. These transnational links are crucial in disseminating the literature, contributing to collaboration patterns and enhanced productivity discussed in Sect. 4.3.

To complete the analysis of the country network dynamic, we include the remaining 403 papers in the dataset to present the current state of the tax avoidance literature in Fig. 4.2. Notably, the literature's expansion in the country coverage has been relatively steady over the past two decades. However, the intensity of connections between countries has experienced a remarkable surge, with the strength of edges increasing more than 15-fold during this period. This indicates a significant growth in the interconnectedness and collaboration among countries within the tax avoidance research domain.

### 4.4.2 *Network Analysis of Affiliations*

In line with the previous analysis, it is no surprise that the United States is the dominant research hub regarding geography. This can be attributed to the significant presence of US-based affiliations, authors, and journals in the field. Our empirical analysis aims to delve deeper into the tax avoidance literature's development and gain a more detailed understanding of where knowledge in this area is generated. To accomplish this, we conduct further research focusing on affiliations.

Importantly, we continue repeating the two steps we have undertaken in the previous analysis. First, we consider the concept of "parent affiliation." The Scopus database records different branches or divisions of the same primary institution as separate entities. For example, SDA Bocconi, Bocconi School of Management, and Bocconi University may have distinct Scopus affiliation IDs. To ensure the accuracy of our analysis, we consolidate these affiliations into their parent institution. This is necessary because the separate listings could create misleading networks, even though these institutions are inherently connected without formal edges. We manually reclassify each author's affiliation by referring to the affiliation hierarchy on the Scopus Affiliation page. However, in this process, we make two exceptions for "The University of North Carolina at Chapel Hill" and "NC State University" to understand their roles in the literature landscape better since the first authors who arguably started the field are or were affiliated with these institutions (e.g., Blouin J., Maydew E., and Shackelford D. A.).

As we examine the data and understand our research objective, we want to integrate a concept that appears necessary in this type of analysis: Unlike a country or an author who is central when it has many connections, an affiliation can be defined as important in a network when it is considered (i.e., cited) by authors belonging to equally important institutions, beyond the number of links itself. As we illustrate in Sect. 3.2.1, this concept can be summarized by the centrality measure of the authority score. So our analysis, as we report from now on, for affiliations is based on this centrality metric, with which we also present some deep dives into the current state of the literature.

Figure 4.6 plots the present literature network of affiliations. The graph reveals a dense, interconnected network of 946 nodes and 22,845 edges. To manage the many nodes, we label only the most central node, *Harvard University*. Table 4.13 summarizes the 40 most central nodes and their centrality degree and authority score measures.

As we examine the data, the previous insight about the centrality of the *United States* is confirmed. The first non-US affiliation, the University of Toronto, occupies the 17th position, while the first European institution, the University of London System, is only at the 26th position.[10] Continuing with American institutions after *Harvard University*, we encounter the Massachusetts Institute of Technology and the University of Chicago. Additionally, prominent tax avoidance literature hubs from the past, such as The University of North Carolina at Chapel Hill, Columbia University, and NC State University, still maintain their status as main hubs today, with relatively high authority scores.

[10] We recall that the University of London System includes both the London Business School and the London School of Economics.

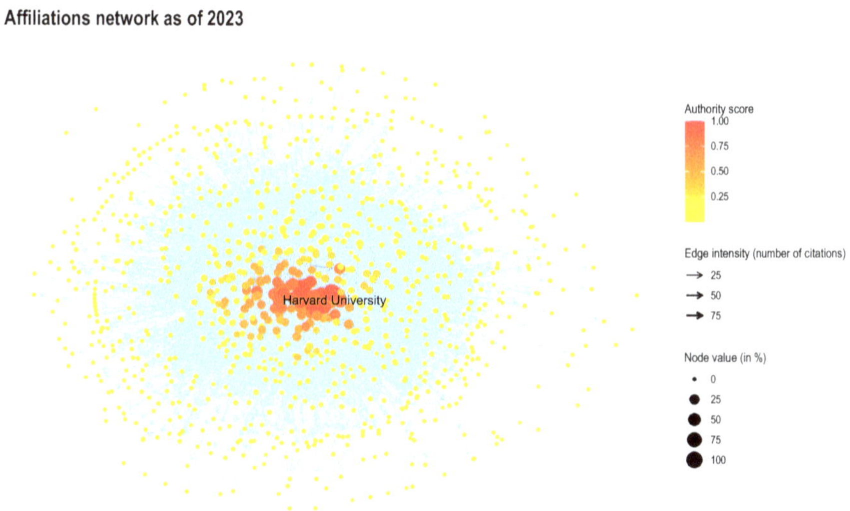


**Fig. 4.6** Affiliations network analysis as of 2023

To gain a more precise visual understanding, we present Figs. 4.7, 4.8, and 4.9, which depict the top 1%, 5%, and 10% affiliations based on their authority scores, respectively. These plots assist us in examining the relationships between significant institutions and the distribution of authority scores. In Fig. 4.7, we observe only ten institutions, all interconnected with strong and reciprocal connections. Moving on to the second plot (Fig. 4.8), there is a significant increase in the number of considered affiliations, resulting in a more concentrated network with a greater number of edges. Among the various connections, we notice thicker lines linking the University of California to both Harvard University and the University of Pennsylvania.

The core network becomes even more connected and interrelated when including all the remaining affiliations to form the top 10%, lacking any meaningful individual connection. This implies a strong interdependence among these nodes. An interesting finding is that, within the top 10%, it is not only universities and research centers that contribute to the literature but also renowned international institutions such as the *World Bank Group* and the *International Monetary Fund*. Similar patterns are evident in previous periods, as discussed below.

#### 4.4.2.1 Time Evolution of Affiliation Network

This section repeats the time evolution analysis for the authors' countries. Examining the initial phase of our literature corpus, we observe 118 nodes and 281 edges. Unlike the country network, the structure here is markedly different. Figure 4.10 illustrates several prominent literature hubs interconnected with satellite institutions right from the outset. The graph highlights the five most central hubs as of 2005:

**Table 4.13** Affiliations network analysis as of 2023 (first 40 rows)

| Node | Affiliation | Degree | Authority | Node | Affiliation | Degree | Authority |
|---|---|---|---|---|---|---|---|
| *150* | Harvard University | 370 | 1.000 | *26* | The University of Arizona | 293 | 0.779 |
| *46* | Massachusetts Institute of Technology | 492 | 0.983 | *60* | Columbia University | 288 | 0.775 |
| *225* | The University of Chicago | 322 | 0.950 | *149* | Indiana University | 299 | 0.772 |
| *121* | University of Pennsylvania | 296 | 0.939 | *76* | The Ohio State University | 270 | 0.768 |
| *65* | The Regents of the University of Michigan | 385 | 0.933 | *206* | University of Rochester | 227 | 0.763 |
| *104* | University of Washington | 316 | 0.929 | *37* | Texas A and M University System | 354 | 0.726 |
| *2* | University of Texas System | 468 | 0.926 | *153* | University of London | 346 | 0.726 |
| *77* | University of California System | 471 | 0.911 | *58* | Arizona State University | 268 | 0.725 |
| *85* | University of Iowa | 377 | 0.907 | *7* | City University of Hong Kong | 416 | 0.725 |
| *42* | Oregon University System | 400 | 0.906 | *56* | New York University | 202 | 0.722 |
| *108* | Duke University | 315 | 0.904 | *347* | University of Colorado System | 185 | 0.693 |
| *187* | Stanford University | 341 | 0.895 | *220* | Dartmouth College | 210 | 0.691 |
| *190* | University of Connecticut | 304 | 0.886 | *100* | Tilburg University | 280 | 0.665 |
| *210* | The University of North Carolina at Chapel Hill | 348 | 0.884 | *1* | Brigham Young University | 242 | 0.655 |
| *80* | University of Wisconsin System | 367 | 0.864 | *6* | Pennsylvania State University | 267 | 0.642 |
| *27* | University of Illinois System | 374 | 0.855 | *45* | Cornell University | 364 | 0.641 |
| *197* | University of Toronto | 293 | 0.815 | *757* | University of Southern California | 150 | 0.632 |
| *5* | State University System of Florida | 451 | 0.796 | *209* | Purdue University System | 204 | 0.629 |
| *312* | Northwestern University | 248 | 0.787 | *212* | Michigan State University | 246 | 0.619 |
| *21* | University System of Georgia | 337 | 0.780 | *14* | NC State University | 238 | 0.597 |

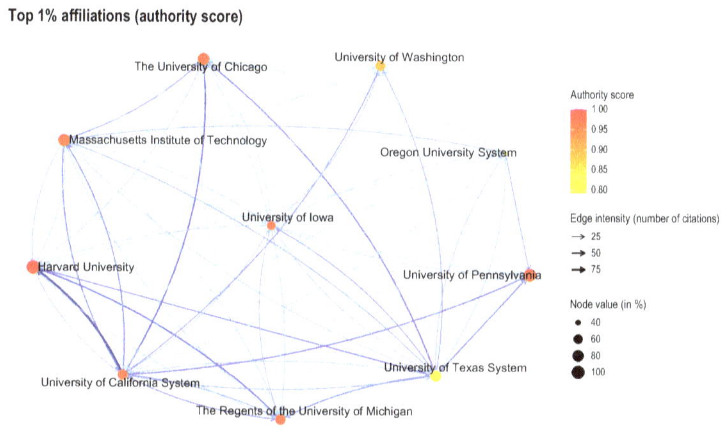


**Fig. 4.7** Top 1% affiliations network analysis as of 2023

*Columbia University*, *The University of North Carolina at Chapel Hill*, *Harvard University*, *The World Bank Group*, and *University of California System*.

In addition to renowned universities, it is crucial to emphasize the significant role international and governmental organizations play in this context. Among the top 40 most central nodes summarized in Table 4.14, we find institutions such as *The World Bank Group*, the *National Bureau of Economic Research*, and the *United States Department of the Treasury*.

Furthermore, the findings align with the literature review and background presented in Chap. 2. For instance, *The University of North Carolina at Chapel Hill* boasts an authority score of 0.938 and a total centrality degree of 11.

Considering the papers published between 2006 and 2010, we find a new affiliations network consisting of 268 nodes and 1080 edges, as shown in Fig. 4.11. Similar to the country analysis, there is a moderate increase in the total number of affiliations compared to the growing number of connections between them (3.84 times). By 2010, the distinct communities observed in the previous graph began diminishing due to multiple connections among the main hubs and surrounding satellite nodes (highlighted as yellow points). American institutions continue to dominate the network's core. Non-US institutions during this period, as indicated in Table 4.15, still occupy lower positions, with *INSEAD* and *Universiteit van Amsterdam* ranking 31st and 32nd, respectively. However, the *Chinese University of Hong Kong* at the 40th position represents an institution from the Eastern world.

As we examine the affiliations network as of 2015, depicted in Fig. 4.12, we observe a total of 448 nodes and 2816 links. Contrary to the previous period, this time frame has contributed more to expanding the overall number of institutions than their interconnections. Consequently, when comparing the shapes of the two networks (Figs. 4.11 and 4.12), they appear quite similar in terms of the central core, with a greater presence of peripheral (yellow) nodes forming an external crown in the 2015 representation. The most central hubs remain unchanged, except for

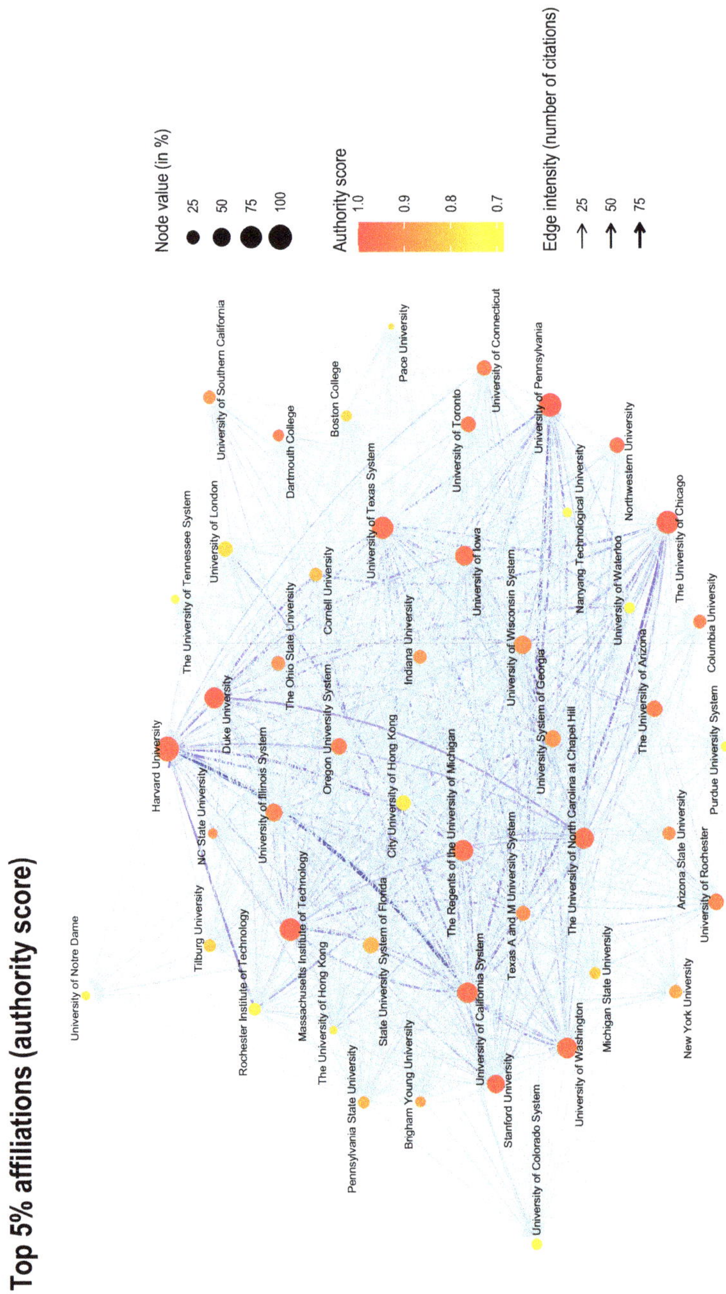


**Fig. 4.8** Top 5% affiliations network analysis as of 2023

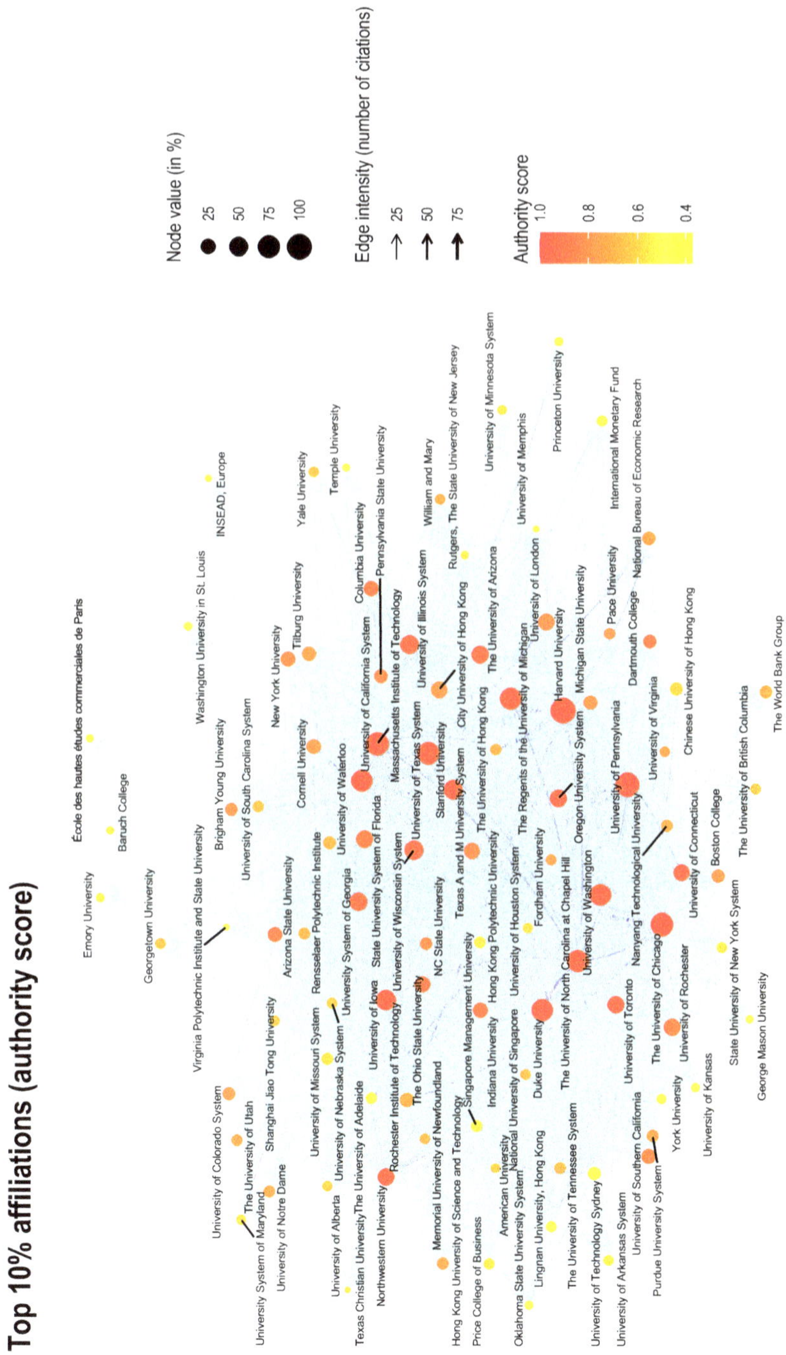


**Fig. 4.9** Top 10% affiliations network analysis as of 2023

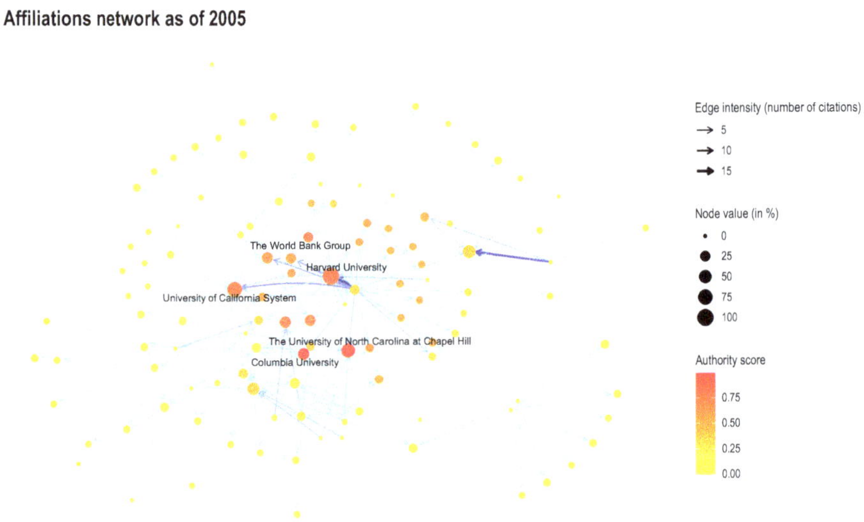


**Fig. 4.10** Affiliations network analysis as of 2005

the *Massachusetts Institute of Technology*, which now holds the second position. Table 4.16 lists the top 40 affiliations based on their authority score as of 2015.

The latest period, from 2016 to the first quarter of 2023, stands out as the most productive, with 403 papers published. When these publications are added to the dataset used for network construction, they contribute to the creation of the graph depicted in Fig. 4.6, which was discussed earlier in this paragraph. As technology advances and new modes of communication emerge, institutions establish closer and more frequent connections. This facilitates the rapid dissemination of knowledge across different realms, mitigating the effects of isolated community "silos." The ongoing evolution observed in countries and affiliations will become even more pronounced in the subsequent paragraphs as we shift the focus to individual authors as the unit of analysis.

### *4.4.3 Network Analysis of Authors*

In the preceding paragraphs, we have extensively explored the intricate connections among various countries and affiliations throughout the analyzed periods. Both network analyses have highlighted the prominent influence of the United States. We now investigate the considerably more intricate relationships between individual authors to complete our comprehensive understanding of the tax avoidance literature. In line with our framework, the authors establish connections through citations, whether they are the ones making citations or receiving them.

**Table 4.14** Affiliations network analysis as of 2005 (first 40 rows)

| Node | Affiliation | Degree | Authority | Node | Affiliation | Degree | Authority |
|---|---|---|---|---|---|---|---|
| *20* | Columbia University | 13 | 1.000 | *58* | Cornell University | 2 | 0.479 |
| *72* | The University of North Carolina at Chapel Hill | 11 | 0.938 | *97* | University of Southern California | 2 | 0.479 |
| *2* | Harvard University | 36 | 0.791 | *107* | University of Delhi | 2 | 0.479 |
| *29* | The World Bank Group | 12 | 0.770 | *112* | Institut d'Économie Industrielle, Toulouse | 2 | 0.479 |
| *4* | University of California System | 29 | 0.712 | *82* | Massachusetts Institute of Technology | 5 | 0.441 |
| *57* | Northwestern University | 6 | 0.711 | *47* | Princeton University | 3 | 0.366 |
| *55* | University of Pennsylvania | 8 | 0.701 | *16* | University of Illinois System | 7 | 0.358 |
| *79* | National Bureau of Economic Research | 5 | 0.612 | *95* | The University of Chicago | 7 | 0.348 |
| *64* | Western University | 3 | 0.570 | *103* | University of Wisconsin System | 3 | 0.340 |
| *76* | Boston University | 3 | 0.570 | *65* | Stanford University | 6 | 0.334 |
| *98* | University of Minnesota System | 4 | 0.570 | *46* | University of London | 4 | 0.324 |
| *69* | United States Department of the Treasury | 4 | 0.551 | *3* | The Regents of the University of Michigan | 36 | 0.303 |
| *75* | Seton Hall University | 3 | 0.541 | *59* | Hebrew University of Jerusalem | 1 | 0.267 |
| *118* | Rutgers, The State University of New Jersey | 4 | 0.507 | *63* | University of Montreal | 1 | 0.267 |
| *62* | University of Massachusetts System | 4 | 0.506 | *91* | Virginia Polytechnic Institute and State University | 1 | 0.267 |
| *32* | University of Oxford | 8 | 0.492 | *110* | Université Fédérale Toulouse Midi-Pyrénées | 1 | 0.267 |
| *41* | State University System of Florida | 5 | 0.479 | *33* | University of Warwick | 3 | 0.267 |
| *44* | Wilfrid Laurier University | 2 | 0.479 | *17* | The University of Arizona | 14 | 0.242 |
| *51* | INSEAD, Europe | 2 | 0.479 | *23* | New York University | 11 | 0.240 |
| *53* | Tel Aviv University | 2 | 0.479 | *43* | University of Colorado System | 7 | 0.207 |

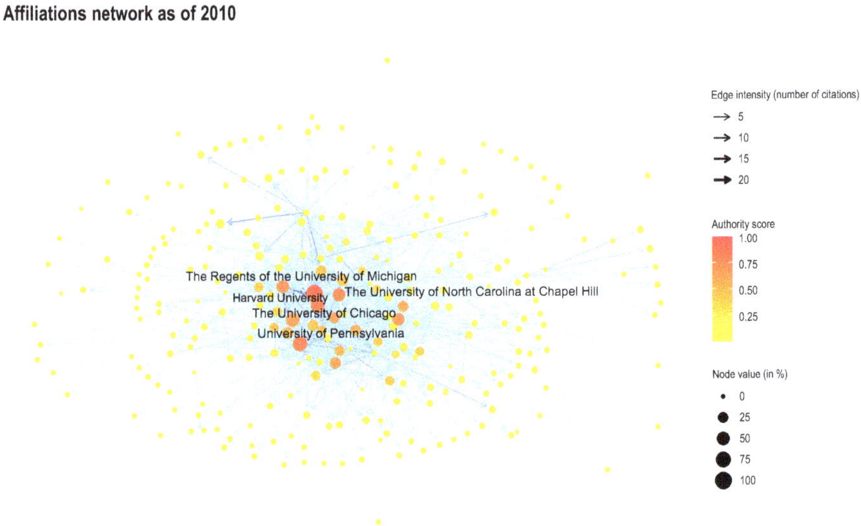


**Fig. 4.11** Affiliations network analysis as of 2010

When comparing the authors' network analysis findings with the previous two analyses, a strikingly different pattern emerges regarding relationships and connections. This network exhibits a completely distinct structure as illustrated in Fig. 4.13. It is not characterized by a central group of highly interconnected nodes surrounded by numerous others interlinked. Instead, we observe a more pronounced presence of distinct author communities within the network. Each node has unique connections that link it to the reference author. While the overall analysis covering all papers from 2000 to the first quarter of 2023 may not exhibit this pattern as clearly, the presence of author communities has consistently been a distinguishing characteristic of these networks over the years, as will be further explored in the analysis of temporal evolution.

In the first quarter of 2023, the author occupying the central position in the network is *Hanlon M.*, affiliated with the *Massachusetts Institute of Technology* (Scopus Author-ID 60022195), with a significant total centrality degree score of 1186. This means 593 authors are directly connected to Hanlon because they have cited her or received citations from her within the papers included in our corpus. The centrality of this author aligns perfectly with the discussions presented in Chap. 2. It is further supported by the centrality of their affiliated institution, as discussed in Sect. 4.4.2. Following closely in the ranking of the most influential authors in the network is *Shevlin T.* (Scopus Author-ID 60116494) from the UCI Paul Merage School of Business (as of the date of analysis), with an impressive count of 917 links connecting him to other authors. This highlights his central role within the academic tax avoidance landscape. The third most influential author, based on the results of our analysis, is *Zuo L.* (Scopus Author-ID 60007776) from Cornell University (as

**Table 4.15** Affiliations network analysis as of 2010 (first 40 rows)

| Node | Affiliation | Degree | Authority | Node | Affiliation | Degree | Authority |
|---|---|---|---|---|---|---|---|
| *9* | Harvard University | 73 | 1.000 | *186* | University of Iowa | 10 | 0.385 |
| *209* | The University of Chicago | 30 | 0.853 | *93* | Yale University | 14 | 0.374 |
| *42* | University of Pennsylvania | 36 | 0.797 | *211* | University of Southern California | 10 | 0.358 |
| *76* | The University of North Carolina at Chapel Hill | 29 | 0.774 | *24* | Pennsylvania State University | 16 | 0.348 |
| *5* | The Regents of the University of Michigan | 84 | 0.714 | *61* | State University System of Florida | 21 | 0.342 |
| *47* | University of Texas System | 40 | 0.699 | *63* | Stanford University | 16 | 0.337 |
| *15* | University of California System | 45 | 0.637 | *127* | The University of British Columbia | 10 | 0.302 |
| *34* | Duke University | 41 | 0.622 | *255* | Rutgers, The State University of New Jersey | 9 | 0.295 |
| *43* | Columbia University | 44 | 0.607 | *37* | City University of Hong Kong | 18 | 0.282 |
| *8* | University of Washington | 40 | 0.577 | *102* | Arizona State University | 9 | 0.274 |
| *198* | University of Rochester | 18 | 0.545 | *103* | INSEAD, Europe | 8 | 0.267 |
| *39* | Massachusetts Institute of Technology | 24 | 0.518 | *54* | Universiteit van Amsterdam | 22 | 0.250 |
| *18* | New York University | 39 | 0.501 | *53* | Michigan State University | 19 | 0.243 |
| *48* | University of Connecticut | 38 | 0.474 | *7* | University of Wisconsin System | 36 | 0.240 |
| *113* | Northwestern University | 14 | 0.457 | *67* | University of Alberta | 16 | 0.229 |
| *46* | The University of Arizona | 20 | 0.436 | *60* | University of Colorado System | 26 | 0.228 |
| *57* | The World Bank Group | 20 | 0.430 | *218* | Syracuse University | 6 | 0.215 |
| *56* | University of Toronto | 19 | 0.414 | *16* | Pace University | 15 | 0.202 |
| *115* | Cornell University | 11 | 0.388 | *130* | Dartmouth College | 5 | 0.202 |
| *171* | National Bureau of Economic Research | 14 | 0.388 | *23* | Chinese University of Hong Kong | 11 | 0.201 |

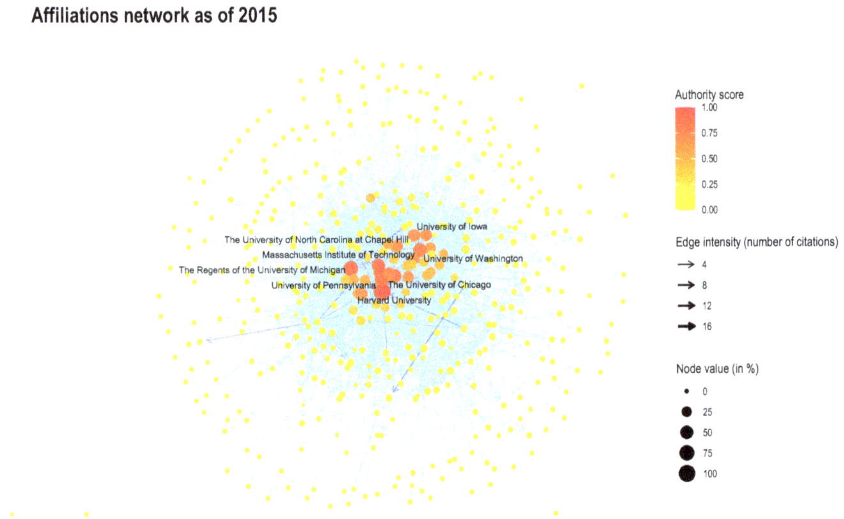


**Fig. 4.12** Affiliations network analysis as of 2015

of the date of analysis). Zuo has contributed significantly to the analyzed corpus, with 1008 citations considered and 98 citations received. (Refer to the example in Sect. 3.3.2 for a detailed understanding of how citations have been counted and classified.) Table 4.17 presents the remaining authors, listed in the first 60 rows, sorted according to their centrality degree scores.

Similarly to the affiliations analysis, we have conducted a detailed examination of the top 1% and 2% of authors ranked by centrality degree to gain a deeper understanding of the underlying structure of this central network. Figure 4.14 clearly illustrates that within the top 1%, there are numerous connections between authors, creating a closely connected central core. This can be observed by the darker blue color that dominates the graph. A comparable pattern is observed in Fig. 4.15, exhibiting a high density of edges among the most influential authors in our analysis.

#### 4.4.3.1 Time Evolution of Authors Network

In the first 5 years of our period, we find that only 292 authors publish tax avoidance-related papers, meeting our publishing journal criteria. Figure 4.16 depicts the connections among these nodes, totaling 625 links. It is essential to highlight the distinct nature of author connections compared to countries and affiliations. Even more evident than in Fig. 4.13, we observe the existence of autonomous and separate author communities during this period. A prominent cluster revolves around the node of *Hines J.* (affiliated with the University of Michigan at that time), with a centrality degree score of 55, indicating 55 arrows that both originate from and point to this node. Notably, no other authors exhibit exceptional centrality degrees (i.e.,

**Table 4.16** Affiliations network analysis as of 2015 (first 40 rows)

| Node | Affiliation | Degree | Authority | Node | Affiliation | Degree | Authority |
|---|---|---|---|---|---|---|---|
| *64* | Harvard University | 122 | 1.000 | *22* | University of Illinois System | 85 | 0.470 |
| *44* | Massachusetts Institute of Technology | 91 | 0.878 | *15* | State University System of Florida | 79 | 0.468 |
| *6* | The Regents of the University of Michigan | 134 | 0.869 | *159* | Arizona State University | 29 | 0.454 |
| *20* | Duke University | 82 | 0.833 | *368* | University of Southern California | 30 | 0.438 |
| *95* | The University of Chicago | 70 | 0.813 | *35* | University System of Georgia | 75 | 0.410 |
| *93* | University of Pennsylvania | 73 | 0.803 | *124* | Cornell University | 40 | 0.380 |
| *28* | University of Washington | 100 | 0.789 | *46* | New York University | 42 | 0.375 |
| *16* | University of Iowa | 69 | 0.714 | *112* | University of Colorado System | 36 | 0.369 |
| *355* | University of Rochester | 49 | 0.704 | *143* | National Bureau of Economic Research | 32 | 0.367 |
| *29* | The University of North Carolina at Chapel Hill | 69 | 0.700 | *52* | Texas A and M University System | 79 | 0.363 |
| *33* | University of Texas System | 105 | 0.689 | *79* | Pace University | 31 | 0.358 |
| *39* | University of Connecticut | 70 | 0.686 | *155* | Indiana University | 32 | 0.349 |
| *14* | University of Wisconsin System | 87 | 0.648 | *67* | Pennsylvania State University | 53 | 0.337 |
| *68* | University of California System | 91 | 0.646 | *178* | Yale University | 22 | 0.323 |
| *1* | The University of Arizona | 78 | 0.588 | *137* | University of London | 37 | 0.310 |
| *212* | Northwestern University | 35 | 0.546 | *40* | Michigan State University | 33 | 0.295 |
| *138* | University of Toronto | 42 | 0.518 | *160* | The World Bank Group | 26 | 0.295 |
| *110* | Oregon University System | 48 | 0.509 | *9* | The Ohio State University | 28 | 0.274 |
| *49* | Columbia University | 64 | 0.506 | *10* | The University of British Columbia | 41 | 0.262 |
| *125* | Stanford University | 54 | 0.490 | *38* | Tilburg University | 77 | 0.262 |

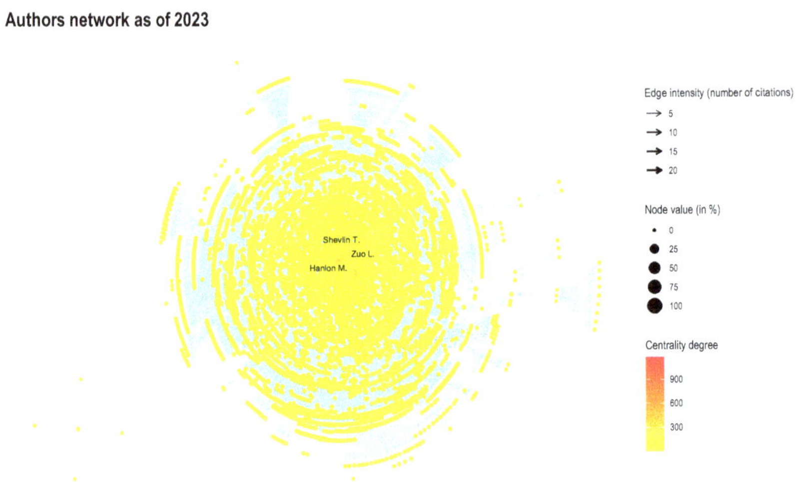


**Fig. 4.13** Authors network analysis as of 2023

**Table 4.17** Authors network analysis as of 2023 (first 60 rows)

| Node | Author | Degree | Node | Author | Degree | Node | Author | Degree |
|---|---|---|---|---|---|---|---|---|
| *67* | Hanlon M. | 1186 | *37* | Lisowsky P. | 395 | *151* | Dhaliwal D.S. | 313 |
| *119* | Shevlin T. | 917 | *387* | Thornock J.R. | 389 | *9* | Kim J.B. | 306 |
| *65* | Zuo L. | 739 | *2597* | Heitzman S. | 373 | *42* | Lobo G.J. | 306 |
| *159* | Wilson R. | 638 | *280* | Lanis R. | 369 | *248* | Hines J. | 305 |
| *388* | Maydew E.L. | 638 | *461* | Zhang L. | 360 | *305* | Hasan M. | 299 |
| *66* | Yeung K. | 636 | *446* | Taylor G. | 345 | *872* | Blouin J. | 285 |
| *127* | Mills L. | 620 | *200* | McGuire S. | 342 | *1759* | Larcker D. | 285 |
| *120* | Klassen K. | 566 | *158* | Wilde J. | 338 | *322* | De Simone L. | 279 |
| *279* | Richardson G. | 545 | *169* | Cheng Q. | 338 | *40* | Donohoe M. | 277 |
| *164* | Wu Q. | 535 | *168* | Chen X. | 336 | *16* | Lewellen C. | 275 |
| *246* | Desai M. | 519 | *542* | Graham J.R. | 331 | *43* | Lee J. | 275 |
| *391* | Dyreng S. | 499 | *110* | Bauer A.M. | 328 | *123* | Gaertner F.B. | 274 |
| *157* | Jacob M. | 497 | *422* | Gallemore J. | 327 | *44* | Lim C.Y. | 271 |
| *620* | Dharmapala D. | 472 | *277* | Watson L. | 325 | *415* | Chyz J.A. | 265 |
| *166* | Zhang H. | 449 | *114* | Hoopes J. | 323 | *4032* | Armstrong C. | 253 |
| *235* | Hasan I. | 443 | *236* | Hoi C. | 323 | *191* | Abdelfattah T. | 246 |
| *308* | Rego S.O. | 441 | *321* | Stomberg B. | 319 | *413* | Schmidt A. | 245 |
| *39* | Pittman J. | 425 | *170* | Chen S. | 316 | *262* | Hardeck I. | 243 |
| *98* | Slemrod J. | 418 | *331* | Laplante S.K. | 316 | *456* | Law K.K.F. | 243 |
| *52* | Omer T. | 415 | *359* | Zhen Li O. | 314 | *479* | Khurana I.K. | 241 |

red shading), but several authors have medium–high degrees, serving as connectors between other individuals and the primary author. These connecting authors are listed in the top rows of Table 4.18.

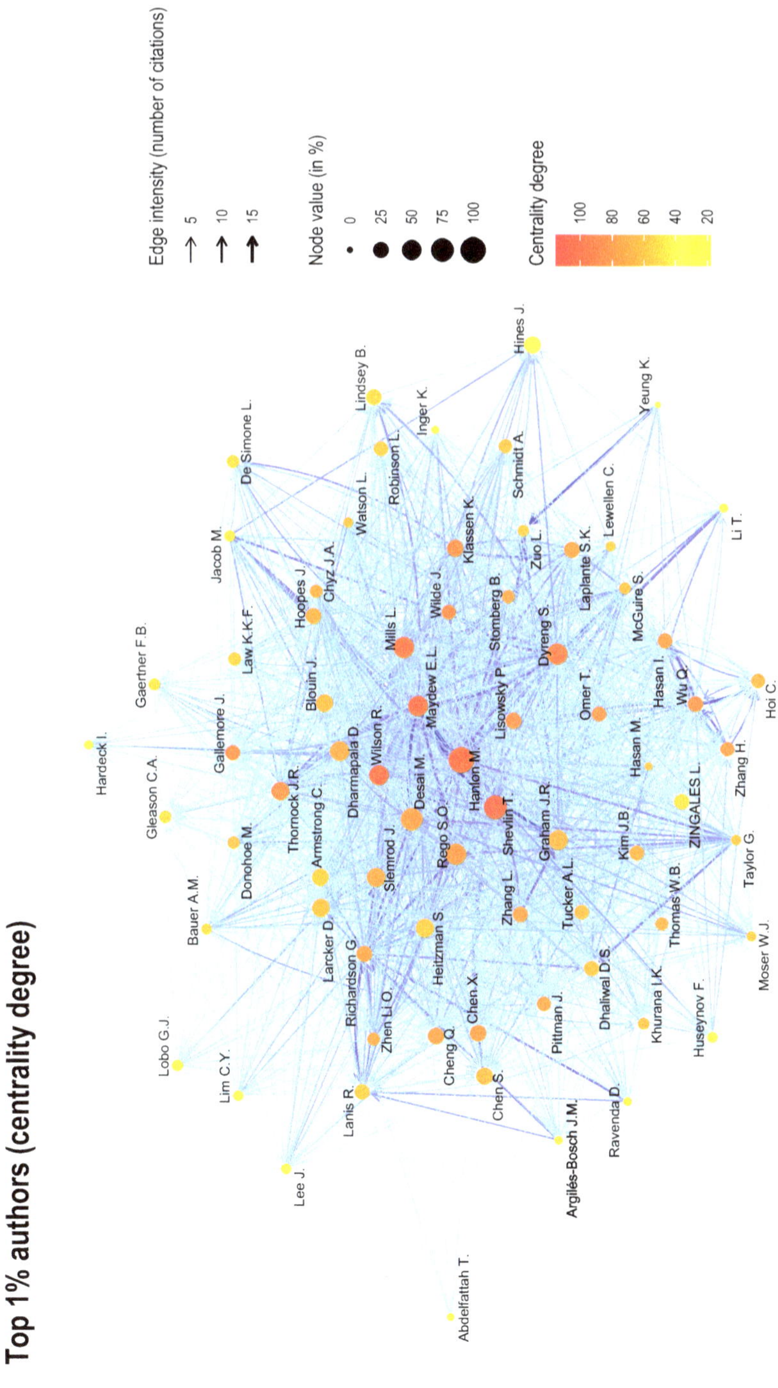


**Fig. 4.14** Top 1% authors network analysis as of 2023 sorted by centrality degree

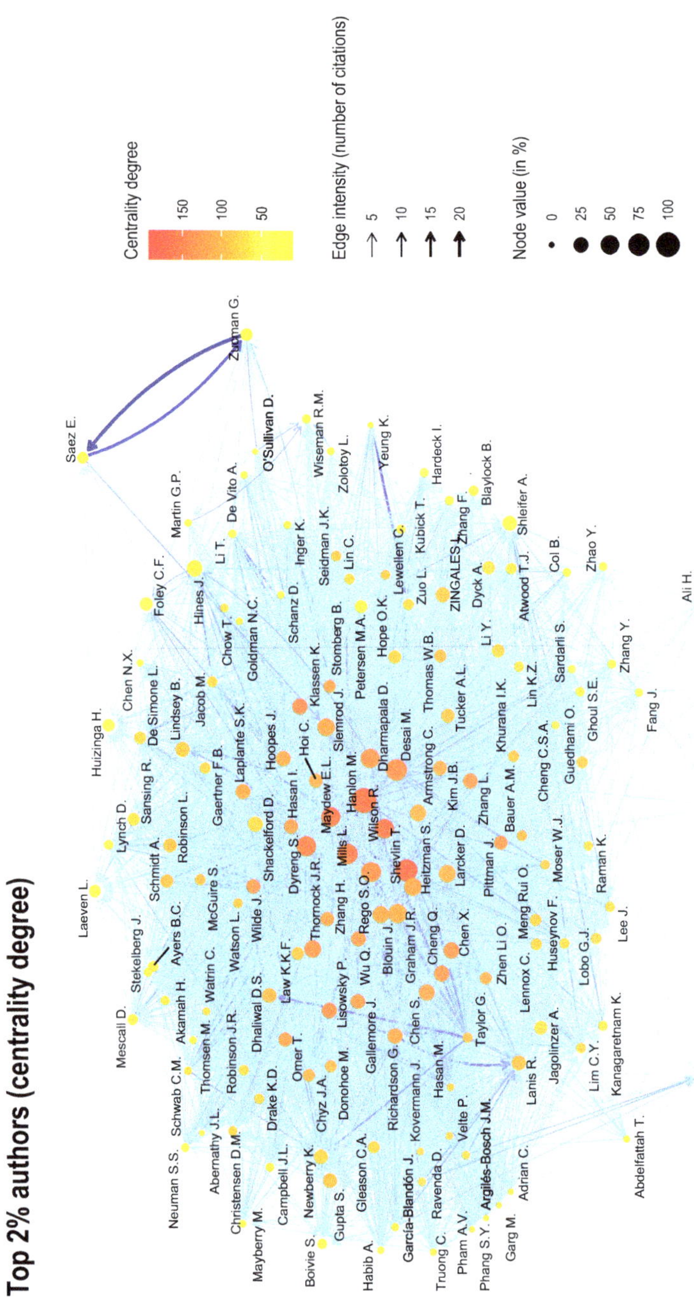


**Fig. 4.15** Top 2% authors network analysis as of 2023 sorted by centrality degree

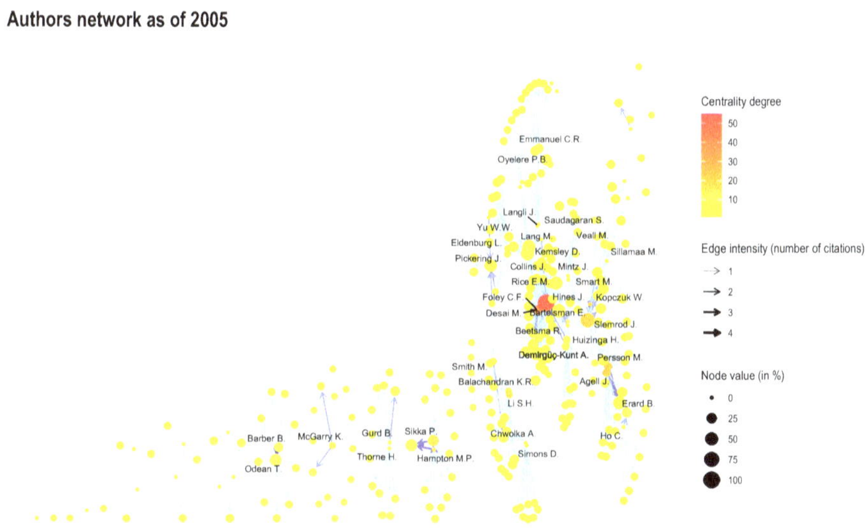


**Fig. 4.16** Authors network analysis as of 2005

In 2010, the results reveal a noticeable growth in the central hub of authors, depicted on the right side of Fig. 4.17, while some small communities still exist outside the core network. The analysis encompasses 960 authors and identifies 3916 links between them. Unlike the trend observed among countries and affiliations, the number of actors among authors demonstrates a relatively higher growth (3.28 times) compared to the increase in links (3.13 times).

This insight encapsulates the essence of our empirical analysis and this monograph, which aims to stimulate collaborations and knowledge exchange among authors, transcending a siloed approach. By doing so, we can harness existing research and build upon it rather than merely generating isolated knowledge.

The foremost author during this period is *Desai M.* from Harvard University, followed by *Dharmapala D.* from the University of Connecticut (as of 2010) and *Hines J.* from the University of Michigan.[11]

Looking at the third snapshot of our research, as summarized in Table 4.20, we observe the well-established central network in 2015 depicted in Fig. 4.18. However, a few groups of 2–3 authors remain somewhat autonomous from the rest of the network. During this period, both the number of authors and the number of connections have experienced significant growth. The number of authors has doubled, reaching 1978 compared to 960 in 2010, while connections increased from 1958 to 4490. Analyzing all the papers from 2000 to 2015, the most influential author emerges as *Desai M.*, followed by *Mills L.*, *Shevlin T.*, *Dhaliwal D. S.*, and *Dharmapala D.* In line with these findings and acknowledging the strong reputation

[11] We refer the reader to Table 4.19 for additional key authors.

**Table 4.18** Authors network analysis as of 2005 (first 60 rows)

| Node | Author | Degree | Node | Author | Degree | Node | Author | Degree |
|---|---|---|---|---|---|---|---|---|
| *4* | Hines J. | 55 | *41* | Eldenburg L. | 13 | *132* | Grubert H. | 9 |
| *2* | Desai M. | 41 | *42* | Pickering J. | 13 | *22* | Cloyd C.B. | 8 |
| *3* | Foley C.F. | 40 | *10* | Emmanuel C.R. | 12 | *48* | Nichols N.B. | 8 |
| *18* | Agell J. | 21 | *11* | Oyelere P.B. | 12 | *49* | Conover T.L. | 8 |
| *19* | Persson M. | 21 | *29* | Erard B. | 12 | *35* | Clausing K. | 7 |
| *5* | Kopczuk W. | 20 | *30* | Ho C. | 12 | *56* | Ward-Batts J. | 7 |
| *6* | Slemrod J. | 19 | *7* | Barber B. | 11 | *57* | Stephens M. | 7 |
| *20* | Sikka P. | 16 | *8* | Odean T. | 11 | *113* | Newberry K. | 6 |
| *21* | Hampton M.P. | 16 | *33* | Beetsma R. | 11 | *115* | Jacob J. | 6 |
| *31* | Balachandran K.R. | 16 | *34* | Bartelsman E. | 11 | *23* | Mills L. | 5 |
| *32* | Li S.H. | 16 | *38* | Huizinga H. | 11 | *84* | Allingham M. | 5 |
| *25* | Chwolka A. | 15 | *39* | Demirgüç-Kunt A. | 11 | *86* | Sandmo A. | 5 |
| *26* | Simons D. | 15 | *51* | Veall M. | 11 | *87* | Sappington D. | 5 |
| *9* | McGarry K. | 14 | *134* | Collins J. | 11 | *95* | Yitzhaki S. | 5 |
| *40* | Yu W.W. | 14 | *135* | Lang M. | 11 | *128* | Shevlin T. | 5 |
| *1* | Smith M. | 13 | *195* | Kemsley D. | 11 | *238* | Alchian A. | 5 |
| *12* | Saudagaran S. | 13 | *36* | Mintz J. | 10 | *24* | Weaver C. | 4 |
| *13* | Langli J. | 13 | *37* | Smart M. | 10 | *47* | Atkinson A.B. | 4 |
| *14* | Gurd B. | 13 | *50* | Sillamaa M. | 10 | *75* | Willmott H. | 4 |
| *15* | Thorne H. | 13 | *215* | Rice E.M. | 10 | *80* | Li L. | 4 |

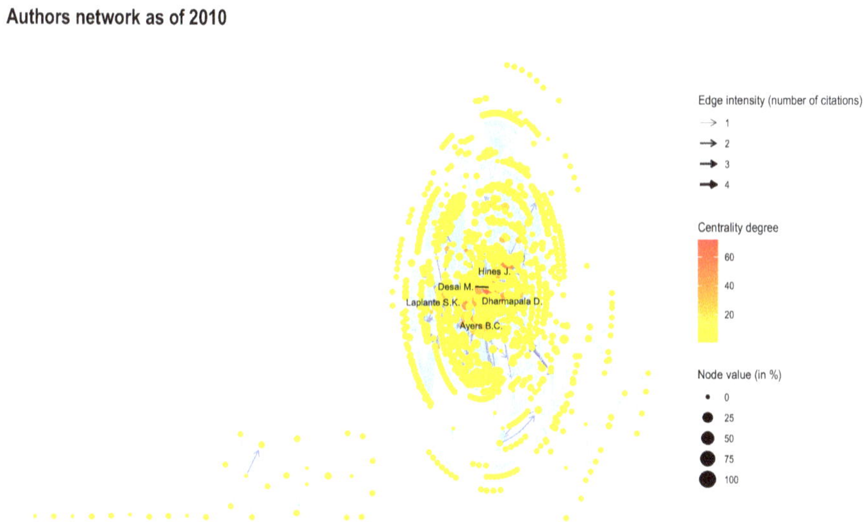


**Fig. 4.17** Authors network analysis as of 2010

of these authors in the tax avoidance field, *Mills L.* and *Shevlin T.* have long served as Editors of *The Accounting Review*. Relatedly, *Desai M.* and *Dharmapala D.* are the authors of two of the most influential papers on corporate governance and tax avoidance (i.e., Desai & Dharmapala, 2006 and Desai & Dharmapala, 2009), whose cumulative citations amount to 4735 (as of 2023).

In the final years of the analyzed period, despite a gradual consolidation, the growth pattern of the authors' network appears distinct from the other entities examined. As mentioned earlier in this section, the authors continue to exhibit organization within reference communities, leading to a greater increase in the number of individual actors compared to the intensity of their relationships.

This authors' network analysis marks the completion of the first and largest part of our empirical research on how the literature on tax avoidance has shaped various connections. In Sect. 4.5, we will conduct a similar comparative analysis from a European perspective. However, we have intentionally kept the study of connections between journals separate to highlight the main outlets publishing influential tax avoidance research. This analysis is discussed in the subsequent section.

### *4.4.4 Network Analysis of Journals*

By examining the connections between countries, affiliations, and authors, we have gained insights into how the literature on tax avoidance has evolved across different geographies and hubs. However, to gain a deeper understanding of the various aspects of this complex and interdisciplinary phenomenon (Hanlon & Heitzman,

**Table 4.19** Authors network analysis as of 2010 (first 60 rows)

| Node | Author | Degree | Node | Author | Degree | Node | Author | Degree |
|---|---|---|---|---|---|---|---|---|
| 10 | Desai M. | 72 | 12 | Krishnamoorthy G. | 33 | 85 | Chen H. | 21 |
| 63 | Dharmapala D. | 66 | 13 | Bedard J. | 33 | 14 | Redoano M. | 19 |
| 11 | Hines J. | 58 | 62 | Baldenius T. | 33 | 41 | Firth M. | 18 |
| 19 | Ayers B.C. | 56 | 91 | Huizinga H. | 32 | 99 | Nicodeme G. | 18 |
| 65 | Laplante S.K. | 52 | 16 | Devereux M.P. | 30 | 4 | Wang H.W. | 17 |
| 26 | Penalva F. | 47 | 24 | Balachandran K.R. | 30 | 33 | Doyle E.M. | 17 |
| 25 | Lara J. | 46 | 3 | Li S.H. | 27 | 34 | Glaister K.W. | 17 |
| 18 | Slemrod J. | 45 | 7 | Chen X. | 26 | 36 | Liu Q. | 17 |
| 68 | Foley C.F. | 43 | 95 | Slangen A. | 26 | 59 | Mills L. | 17 |
| 1 | Omer T. | 41 | 35 | Sikka P. | 25 | 74 | Dyreng S. | 17 |
| 44 | Graham J.R. | 40 | 77 | Jiang J. | 25 | 84 | Onji K. | 17 |
| 46 | Kim B. | 40 | 23 | Wang T. | 24 | 98 | Morck R. | 17 |
| 47 | Kim B. | 39 | 96 | Smeets R. | 24 | 17 | Kopczuk W. | 16 |
| 22 | Laeven L. | 38 | 2 | Falsetta D. | 23 | 27 | Azémar C. | 16 |
| 43 | Osma B. | 37 | 60 | Cheng Q. | 23 | 89 | Reichelstein S. | 16 |
| 8 | Shevlin T. | 36 | 61 | Chen S. | 23 | 568 | Maydew E.L. | 16 |
| 37 | Beugelsdijk S. | 36 | 39 | Willmott H. | 22 | 48 | Lo A. | 15 |
| 45 | Jung K. | 36 | 9 | Hanlon M. | 21 | 49 | Wong R. | 15 |
| 64 | McGuire S. | 36 | 15 | Lockwood B. | 21 | 330 | Newberry K. | 15 |
| 38 | Hennart J.F. | 34 | 21 | Tucker A.L. | 21 | 755 | Rice E.M. | 15 |

**Table 4.20** Authors network analysis as of 2015 (first 60 rows)

| Node | Author | Degree | Node | Author | Degree | Node | Author | Degree |
|---|---|---|---|---|---|---|---|---|
| *87* | Desai M. | 110 | *46* | Wu Q. | 47 | *134* | Robin A.J. | 36 |
| *40* | Mills L. | 107 | *77* | Sikka P. | 47 | *4* | Gavious I. | 35 |
| *85* | Shevlin T. | 104 | *26* | Richardson G. | 45 | *66* | Penalva F. | 35 |
| *55* | Dhaliwal D.S. | 84 | *59* | Lanis R. | 44 | *10* | Tang T.Y.H. | 34 |
| *124* | Dharmapala D. | 84 | *181* | Foley C.F. | 44 | *56* | Myllymäki E.R. | 34 |
| *151* | Laplante S.K. | 81 | *186* | Dyreng S. | 44 | *61* | Huizinga H. | 34 |
| *17* | Wilson R. | 73 | *118* | Law K.K.F. | 43 | *155* | Meng Rui O. | 34 |
| *38* | Slemrod J. | 73 | *189* | Lennox C. | 42 | *19* | Yosef R. | 33 |
| *58* | Graham J.R. | 70 | *2* | Pittman J. | 41 | *111* | Tucker A.L. | 33 |
| *144* | Hanlon M. | 70 | *43* | Ayers B.C. | 41 | *119* | Taylor G. | 33 |
| *25* | Hines J. | 63 | *174* | Schmidt A. | 41 | *154* | Chyz J.A. | 32 |
| *33* | Maydew E.L. | 62 | *86* | Graffin S.D. | 40 | *139* | Sugathan A. | 31 |
| *72* | McGuire S. | 61 | *92* | Boivie S. | 40 | *74* | Doyle E.M. | 30 |
| *54* | Omer T. | 60 | *150* | Klassen K. | 40 | *105* | Chen E. | 30 |
| *278* | Rego S.O. | 53 | *91* | Christensen D.M. | 38 | *164* | Baldenius T. | 30 |
| *57* | Zhang H. | 51 | *148* | Chen X. | 38 | *914* | Shackelford D. | 30 |
| *152* | Khurana I.K. | 51 | *149* | Chen S. | 38 | *29* | Abdul Wahab N. | 28 |
| *24* | Lisowsky P. | 48 | *32* | Thornock J.R. | 37 | *36* | Zhen Li O. | 28 |
| *49* | Donohoe M. | 48 | *35* | Robinson L. | 37 | *52* | Ravenda D. | 28 |
| *44* | Hasan I. | 47 | *15* | Cheng Q. | 36 | *70* | Osma B. | 28 |

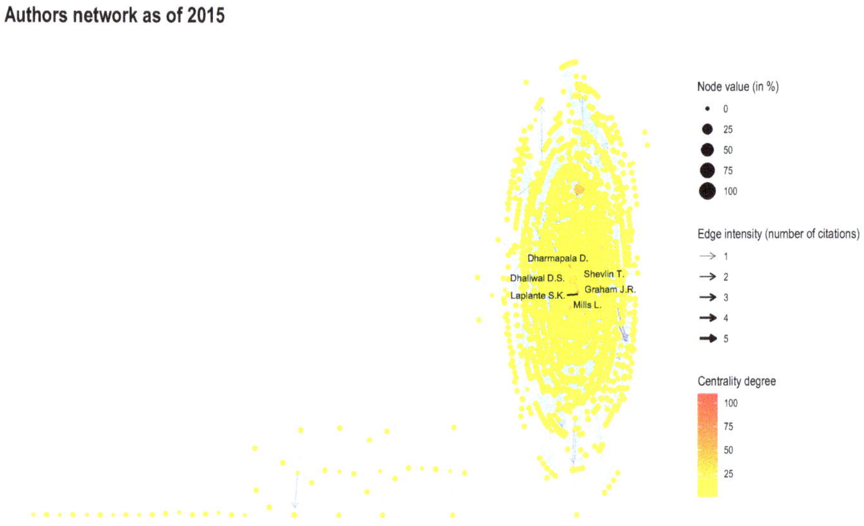


Fig. 4.18 Authors network analysis as of 2015

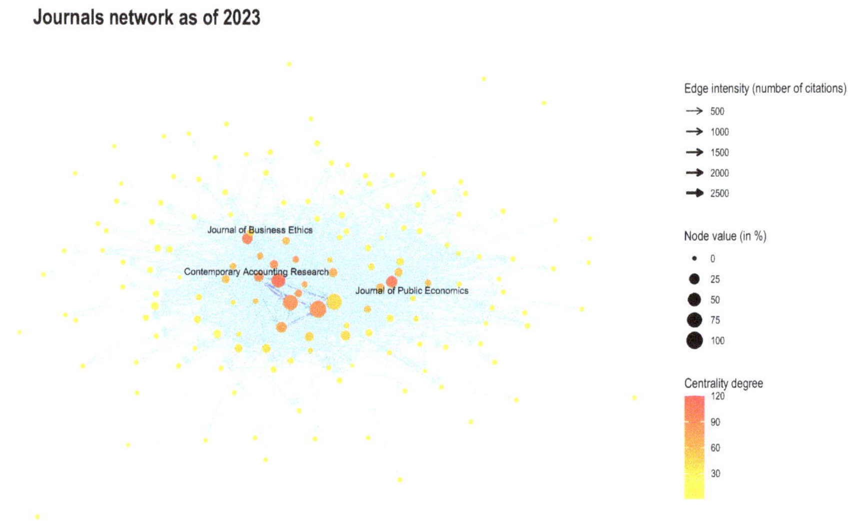


Fig. 4.19 Journals network analysis as of 2023

2010), we must now turn our attention to the analysis of journals. This will provide us with a clearer picture of which specific aspects of tax avoidance have been the focal points at different times.

Upon examining Fig. 4.19, it becomes apparent that the publishers in our analysis are interconnected, forming a network. However, what truly captures our attention is identifying the most influential journals within this network. Table 4.21 reveals that

**Table 4.21** Journals network analysis as of 2023 (first 20 rows)

| Node | Journal | ABS | Degree |
|---|---|---|---|
| *10* | Contemporary Accounting Research | 4 | 120 |
| *18* | Journal of Public Economics | 3 | 109 |
| *2* | Journal of Business Ethics | 3 | 101 |
| *8* | Journal of the American Taxation Association | 3 | 94 |
| *9* | Journal of Accounting Research | 4* | 86 |
| *25* | Journal of Financial Economics | 4* | 86 |
| *23* | Accounting Review | 4* | 85 |
| *3* | European Accounting Review | 3 | 83 |
| *24* | Accounting and Business Research | 3 | 83 |
| *5* | Review of Accounting Studies | 4 | 74 |
| *1* | Journal of International Business Studies | 4* | 68 |
| *7* | Journal of Business Finance and Accounting | 3 | 68 |
| *27* | Journal of International Accounting, Auditing and Taxation | 3 | 66 |
| *44* | Quarterly Journal of Economics | 4* | 63 |
| *15* | Accounting Forum | 3 | 59 |
| *20* | Journal of Financial and Quantitative Analysis | 4 | 59 |
| *22* | American Economic Journal: Economic Policy | 3 | 58 |
| *17* | Strategic Management Journal | 4* | 56 |
| *48* | Review of Economics and Statistics | 4 | 56 |
| *62* | American Economic Review | 4* | 55 |

*Contemporary Accounting Research* holds the top position with a centrality degree score of 120, followed by the *Journal of Public Economics* (degree 109) and the *Journal of Business Ethics* (degree 101). These findings lead us to two interesting conclusions.

First, from an analytical standpoint, none of the mentioned journals are rated as 4* in the ABS journal list, indicating that the centrality of a journal does not necessarily correlate with its ranking. Second, and more importantly for our research, if we consider the centrality of a journal as an indicator of the relative importance of a topic within the field of tax avoidance, we can assert that the most frequently discussed aspects revolve around accounting, as demonstrated by the prominence of *Contemporary Accounting Research* in the literature. This observation aligns seamlessly with the extensive discussion presented in Chap. 2, where we qualitatively traced the evolution of the literature over the years, emphasizing the growing significance of analyzing the determinants and the real effects of tax avoidance.

#### 4.4.4.1 Time Evolution of Journals Network

The evolution of the journal network throughout the years has demonstrated a distinct pattern of progressive consolidation, diverging from that observed in countries, affiliations, and authors. While the former cases have exhibited relatively slow growth in new nodes compared to the accelerated intensification of links, juxtaposed with a rapid rise in the number of new authors compared to their interconnections, the growth pattern of journals can be described as hybrid.

Figure 4.20 shows that distinct journal communities with central and satellite nodes become apparent when examining the papers from the initial 5-year period. Notably, the *Journal of Public Economics*, *Journal of Financial Economics*, and *European Accounting Review* exemplify this pattern. These journals also occupy top positions in Table 4.22, highlighting the 20 most central nodes. This observation aligns with our qualitative analysis, emphasizing the substantial growth of the literature during these initial years, mainly focusing on the determinants of tax avoidance within the accounting and finance realms.

Including 30 additional journals that published papers from our dataset between 2006 and 2010 expands the number of nodes considered in Fig. 4.21 to 87, interconnected by 277 links. Upon inspection, we can discern a process of consolidation among the previously distinct communities observed in the initial snapshot. It is worth noting that while there continues to be a prevalence of accounting-focused journals, the *Journal of Public Economics* retains its position as the most central journal. Common to other economics journals (e.g., Journal of International Eco-

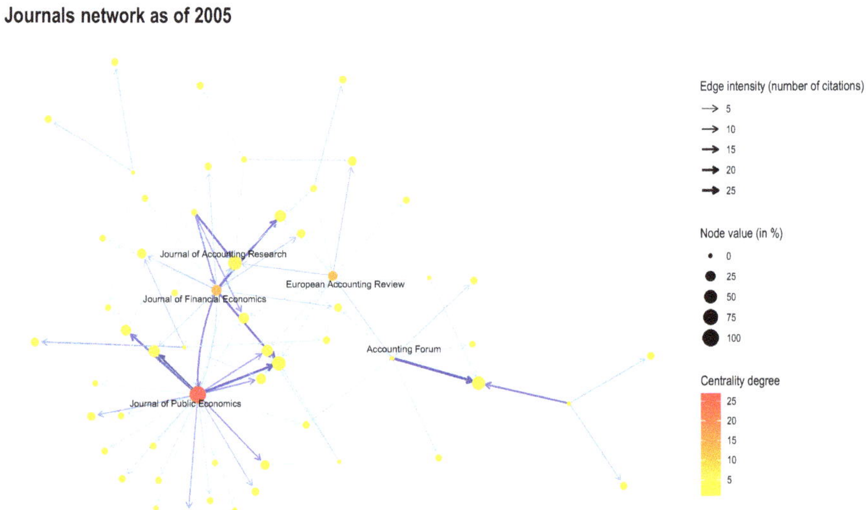


**Fig. 4.20** Journals network analysis as of 2005

**Table 4.22** Journals network analysis as of 2005 (first 20 rows)

| Node | Journal | ABS | Degree |
|---|---|---|---|
| *3* | Journal of Public Economics | 3 | 27 |
| *2* | Journal of Financial Economics | 4* | 14 |
| *5* | European Accounting Review | 3 | 13 |
| *8* | Accounting Forum | 3 | 7 |
| *13* | Journal of Accounting Research | 4* | 7 |
| *10* | Review of Quantitative Finance and Accounting | 3 | 6 |
| *11* | International Journal of Accounting | 3 | 5 |
| *1* | Journal of Accounting, Auditing and Finance | 3 | 4 |
| *4* | Journal of International Accounting, Auditing and Taxation | 3 | 4 |
| *9* | Journal of the American Taxation Association | 3 | 4 |
| *15* | American Economic Review | 4* | 4 |
| *32* | Quarterly Journal of Economics | 4* | 4 |
| *33* | Journal of Accounting and Economics | 4* | 4 |
| *6* | Financial Accountability and Management | 3 | 3 |
| *12* | Journal of the European Economic Association | 4 | 3 |
| *14* | Journal of International Business Studies | 4* | 3 |
| *19* | Accounting, Organizations and Society | 4* | 3 |
| *55* | Econometrica | 4* | 3 |
| *7* | Abacus | 3 | 2 |
| *23* | Journal of Economic Behavior and Organization | 3 | 2 |

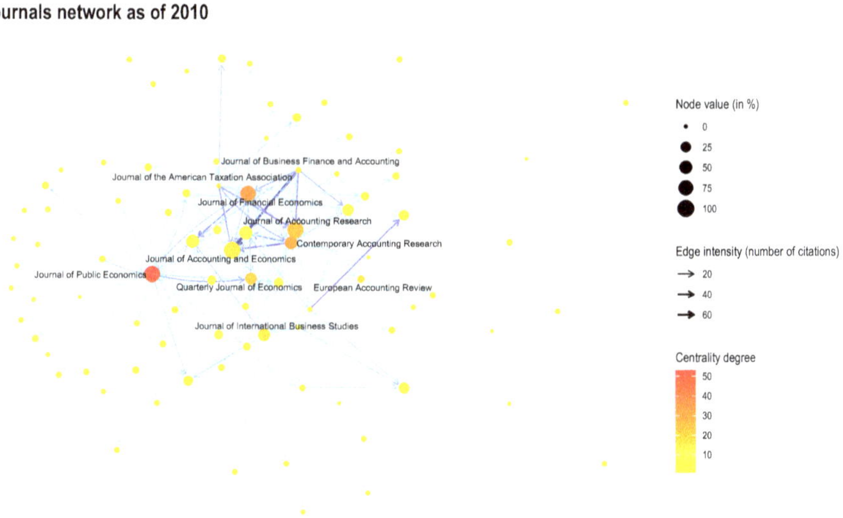


**Fig. 4.21** Journals network analysis as of 2010

**Table 4.23** Journals network analysis as of 2010 (first 20 rows)

| Node | Journal | ABS | Degree |
|---|---|---|---|
| *3* | Journal of Public Economics | 3 | 53 |
| *4* | Journal of Financial Economics | 4* | 33 |
| *5* | Contemporary Accounting Research | 4 | 29 |
| *22* | Journal of Accounting Research | 4* | 23 |
| *21* | Quarterly Journal of Economics | 4* | 20 |
| *1* | Journal of the American Taxation Association | 3 | 17 |
| *6* | Journal of Business Finance and Accounting | 3 | 16 |
| *16* | European Accounting Review | 3 | 16 |
| *13* | Journal of International Business Studies | 4* | 15 |
| *42* | Journal of Accounting and Economics | 4* | 15 |
| *11* | Economic Journal | 4 | 14 |
| *2* | Journal of Accounting, Auditing and Finance | 3 | 13 |
| *10* | Accounting Forum | 3 | 13 |
| *12* | Review of Economics and Statistics | 4 | 13 |
| *14* | Critical Perspectives on Accounting | 3 | 11 |
| *23* | Foundations and Trends in Accounting | 3 | 11 |
| *48* | Accounting Review | 4* | 10 |
| *31* | American Economic Review | 4* | 9 |
| *41* | Review of Accounting Studies | 4 | 9 |
| *63* | Econometrica | 4* | 9 |

nomics), this journal is a significant reference point in the literature to understand *how* firms engage in tax avoidance (i.e., through income shifting).

Another noteworthy aspect, as mentioned earlier, pertains to the ABS journal ranking. Among the top 20 entries in Table 4.23, we observe that only eight journals are ranked as 4*, four have a ranking of 4, and the remaining ones are classified as ABS 3. This indicates that tax avoidance has garnered increasing interest over time, even from prestigious journals, while also finding ample publishing opportunities in more specialized journals with ABS 3 rankings.

Concluding our analysis of the journal network over time, we incorporate the journals that published papers between 2011 and 2015, aligning with the framework discussed earlier in this section. Figure 4.22 presents a consolidated network, showcasing connections between the central nodes and among the peripheral nodes, depicted in yellow, which were previously linked to the core solely through a single connection with their respective reference node.

Examining the individual journals in greater detail, as presented in Table 4.24, we still observe a significant dominance of accounting journals even with the inclusion of an additional 5-year period (2000–2015) in the analysis. However, their importance is slightly diminished compared to the *Journal of Public Economics*, which maintains the highest degree score of 71. This finding suggests that tax accounting scholars may also find economics outlets suitable for publishing their

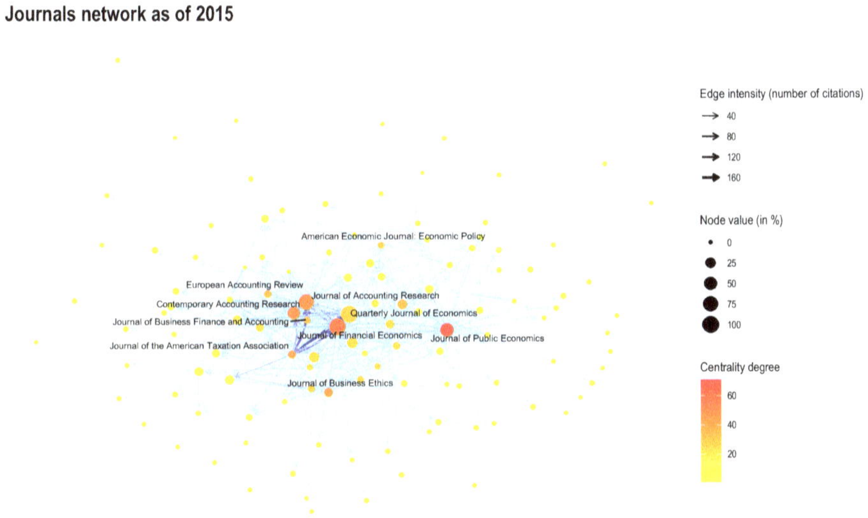


**Fig. 4.22** Journals network analysis as of 2015

**Table 4.24** Journals network analysis as of 2015 (first 20 rows)

| Node | Journal | ABS | Degree |
|---|---|---|---|
| *4* | Journal of Public Economics | 3 | 71 |
| *9* | Journal of Financial Economics | 4* | 57 |
| *2* | Journal of Accounting Research | 4* | 48 |
| *14* | Contemporary Accounting Research | 4 | 48 |
| *1* | Journal of the American Taxation Association | 3 | 43 |
| *11* | Journal of Business Ethics | 3 | 42 |
| *5* | European Accounting Review | 3 | 32 |
| *19* | Journal of Business Finance and Accounting | 3 | 32 |
| *20* | Quarterly Journal of Economics | 4* | 32 |
| *6* | American Economic Journal: Economic Policy | 3 | 27 |
| *8* | Critical Perspectives on Accounting | 3 | 25 |
| *15* | Accounting and Business Research | 3 | 25 |
| *48* | Journal of Accounting and Economics | 4* | 25 |
| *23* | Accounting Forum | 3 | 24 |
| *13* | Journal of Finance | 4* | 23 |
| *16* | Strategic Management Journal | 4* | 23 |
| *18* | Journal of International Accounting, Auditing and Taxation | 3 | 23 |
| *25* | Journal of International Business Studies | 4* | 23 |
| *33* | Review of Economics and Statistics | 4 | 23 |
| *22* | Accounting Review | 4* | 22 |

tax avoidance research. Additionally, the *Journal of Business Ethics*, which focuses on social and ethical issues as it is tax avoidance, emerges among the top 20 journals for centrality. As previously discussed in the comprehensive analysis, this journal assumes a position of absolute centrality in the 2023 network.

Completing this final analysis, our empirical research on the international tax avoidance literature ends. In the subsequent section, we will focus on conducting a similar analysis exclusively on European countries.[12] The aim is to assess any similarities and differences, in terms of both the most central entities and the temporal progression of the networks, within the context of Europe.

## 4.5 The European Network

Based on our empirical analysis of the tax avoidance research, we have obtained valuable insights that support and expand upon the literature review conducted earlier. Specifically, our examination of the various networks has revealed a gradual convergence between countries and affiliations, with increasing interactions occurring over time. This convergence has led to forming a closely interconnected group of actors. Notably, the central role of the United States was evident in both country and affiliation networks. In contrast, we have observed a significant growth in the number of scholars dedicated to studying tax avoidance over the past two decades. However, the connections between these authors still appear to have ample room for further expansion and development.

The importance of focusing on Europe in our research stems from the need to address a specific question that complements the global perspective analyzed earlier in this monograph. While the previous analysis has shed light on the origins and dissemination of the tax avoidance literature on a global scale, our examination of European countries seeks to answer a different question. Specifically, we aim to identify the primary reference points available to European scholars studying tax avoidance and investigate how these references have evolved. By focusing on Europe, we can gain insights into the specific scholarly landscape within this region, understanding the prominent actors, affiliations, and countries that contribute to the tax avoidance literature. This analysis allows us to discern the trajectory and transformations of research in the European context, highlighting the influential sources and networks that European scholars can draw upon. Furthermore, by comparing these findings with the global analysis, we can identify similarities and differences, uncovering European scholarship's unique contributions and dynamics on tax avoidance.

To address the question, we employ the previous approach and analyze networks of countries, affiliations, and authors within the European context. However, before

[12] In the analysis, we refer to "European countries," "Europe," or "EU27 Countries" interchangeably.

delving into the results of our analysis, we will provide a brief overview of the initial steps taken in filtering the dataset using the software R. Subsequently, we will present our findings, beginning with an examination of countries, followed by an investigation of affiliations and authors. Finally, we will conclude the analysis by exploring the networks of journals, which offer unique insights into the prominence of specific topics within the broader domain of tax avoidance. By exploring these different facets, we can comprehensively understand the intricate dynamics and central themes within tax avoidance research.

### 4.5.1 Data Cleaning and Alignment

To adopt a Europe-centric approach, we initially filter our corpus based on the location of the parent affiliation of the paper's author. Before delving into the specifics of our analysis, it is important to clarify two methodological aspects:

1. The European filter is exclusively applied to the papers and not the references cited. We deem it impractical to eliminate citations from outside the EU, as they are expected to constitute the majority. Removing them would significantly reduce the sample size and could lead to inaccurate analysis and potentially flawed results.
2. The centrality measure employed in all analyses is the centrality degree score. Building upon the insights gained from our previous analysis, which has highlighted the United States as the central hub of tax avoidance research, our objective in this second analysis is to identify the nodes with the highest degree of connectivity. This information will be valuable for European researchers seeking to enter the global network and engage with the most influential players.

These clarifications provide a solid foundation as we delve into the subsequent analysis, enabling us to gain a comprehensive understanding of the European landscape in tax avoidance research. The applied filter consists of the list of European countries, as indicated in the code lines shown in Fig. 4.23. As a result, the number of papers considered within this new perspective decreases to 160.

### 4.5.2 Network Analysis of European Countries

Continuing with the methodology employed earlier, we initiate an in-depth examination of the European Area countries by investigating their interconnections, as depicted in Fig. 4.24.[13]

[13] The European Area is a zone comprising 27 European countries that have abolished internal border controls, allowing for the free movement of people within the area. It was established through the 1985 Schengen Agreement and the 1990 Schengen Convention.

```
# DATA FILTERING ---------------------------------------------------------------
# We use Schengen Area as a filter
colnames(network_dataset)
eu_network_dataset <-
  network_dataset[network_dataset$`paper-author-affilcountry` == "Austria" |
                  network_dataset$`paper-author-affilcountry` == "Belgium" |
                  network_dataset$`paper-author-affilcountry` == "Czech Republic" |
                  network_dataset$`paper-author-affilcountry` == "Denmark" |
                  network_dataset$`paper-author-affilcountry` == "Finland" |
                  network_dataset$`paper-author-affilcountry` == "France" |
                  network_dataset$`paper-author-affilcountry` == "Germany" |
                  network_dataset$`paper-author-affilcountry` == "Greece" |
                  network_dataset$`paper-author-affilcountry` == "Hungary" |
                  network_dataset$`paper-author-affilcountry` == "Iceland" |
                  network_dataset$`paper-author-affilcountry` == "Italy" |
                  network_dataset$`paper-author-affilcountry` == "Luxembourg" |
                  network_dataset$`paper-author-affilcountry` == "Netherlands" |
                  network_dataset$`paper-author-affilcountry` == "Norway" |
                  network_dataset$`paper-author-affilcountry` == "Portugal" |
                  network_dataset$`paper-author-affilcountry` == "Slovakia" |
                  network_dataset$`paper-author-affilcountry` == "Spain" |
                  network_dataset$`paper-author-affilcountry` == "Sweden" |
                  network_dataset$`paper-author-affilcountry` == "Switzerland", ]
```

**Fig. 4.23** The code to filter our network dataset considering only European countries

In the center of this network, nodes colored in various shades of red represent countries such as the Netherlands, Germany, France, and Finland, which exhibit the highest number of connections within this network (i.e., they have a higher centrality degree, as indicated in Table 4.25). The darker blue arrows highlight the predominant connections directed toward the United States. This observation indicates many citations from authors affiliated with US institutions by German, Dutch, French, and Finnish authors within our network.[14] The graph also includes non-EU countries, although they do not serve as the polarizing center of the network. Instead, these countries serve as sources from which European scholarship derives information and previous knowledge. These non-EU countries are associated with papers written by scholars whose affiliations are located outside of Europe and have been cited by the collection of authors within our sample.

#### 4.5.2.1 Time Evolution of European Countries Network

When focusing solely on European papers, the analysis of the temporal evolution of connections reveals smaller networks in terms of both the number of nodes (i.e., countries) and the number of connections between them.

[14] In our network, a link from node *A* to node *B* implies that node *B* has been cited by node *A*. Hence, we observe numerous citations from authors affiliated with US institutions by German, Dutch, French, and Finnish authors.

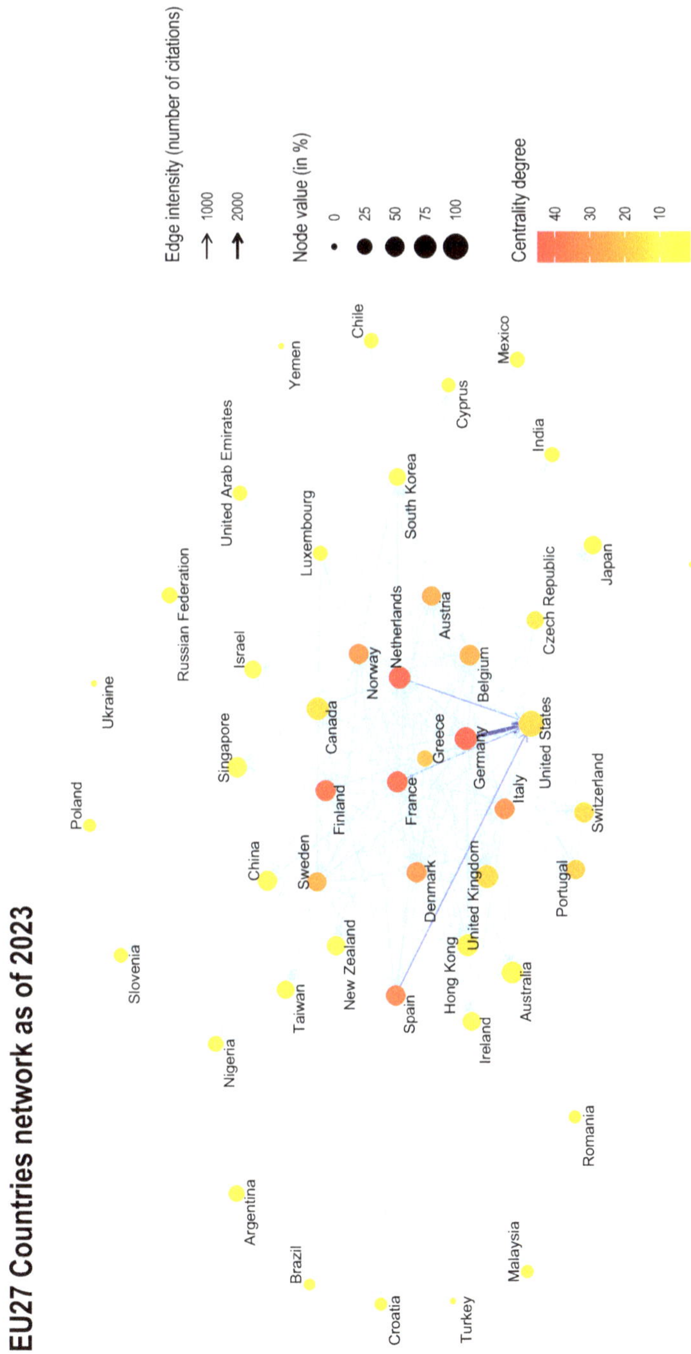


**Fig. 4.24** European countries network analysis as of 2023

**Table 4.25** European countries network analysis as of 2023 (first 30 rows)

| Node | Country | Degree |
|---|---|---|
| *2* | Netherlands | 45 |
| *6* | Germany | 45 |
| *8* | France | 42 |
| *11* | Finland | 40 |
| *5* | Spain | 36 |
| *12* | Italy | 32 |
| *1* | Norway | 30 |
| *7* | Denmark | 30 |
| *4* | Austria | 26 |
| *14* | Belgium | 25 |
| *3* | Sweden | 24 |
| *9* | Greece | 21 |
| *10* | Portugal | 18 |
| *16* | Switzerland | 16 |
| *17* | United States | 16 |
| *19* | United Kingdom | 16 |
| *18* | Canada | 15 |
| *13* | Czech Republic | 14 |
| *22* | Hong Kong | 13 |
| *15* | Luxembourg | 11 |
| *34* | Australia | 11 |
| *20* | New Zealand | 9 |
| *23* | South Korea | 9 |
| *27* | Singapore | 9 |
| *31* | China | 9 |
| *25* | Ireland | 8 |
| *28* | Israel | 8 |
| *26* | Taiwan | 6 |
| *35* | Japan | 6 |
| *32* | United Arab Emirates | 4 |

Examining the first period under analysis, which encompasses studies published between 2000 and 2005, Fig. 4.25 displays only 11 nodes (including five non-European countries, such as the United Kingdom) connected by a mere 15 connections. As discussed in Sect. 4.5.2, the connections between *Germany* and the *United States*, as well as between *Norway* and the *United States*, are significantly robust. These two European countries, along with *Sweden* (which, however, has connections beyond the United States), exhibit a higher centrality degree of 5, as summarized in Table 4.26.

This analysis allows us to draw a conclusion that aligns with the discussions in Chap. 2 and the previous empirical analysis: The development of tax avoidance research in Europe has been slower compared to Anglo-Saxon countries, particularly the United States and Canada. European research in this field has relied on

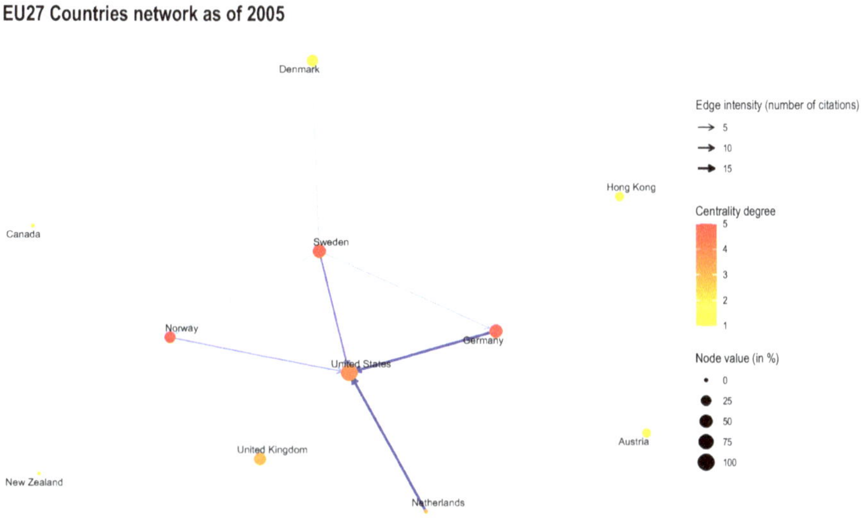


**Fig. 4.25** European countries network analysis as of 2005

**Table 4.26** European countries network analysis as of 2005

| Node | Country | Degree |
|---|---|---|
| *1* | Norway | 5 |
| *3* | Sweden | 5 |
| *4* | Germany | 5 |
| *6* | United States | 4 |
| *2* | Netherlands | 3 |
| *5* | United Kingdom | 3 |
| *7* | New Zealand | 1 |
| *8* | Hong Kong | 1 |
| *9* | Canada | 1 |
| *10* | Denmark | 1 |
| *11* | Austria | 1 |

the foundations laid by these two countries. However, a slightly different outcome emerges when we consider all the papers published between 2006 and 2010 in addition to the previous ones. Figure 4.26 reveals a network of nodes that begins to take shape, with a greater number of nodes (18) and more interconnections (36 edges). Countries like *Spain*, *France*, and *Belgium* are starting to play significant roles and foster connections (as evidenced by the strong link between the United States and Spain). Nevertheless, *Germany* and *Sweden* remain the primary countries, as indicated in Table 4.27, with the *Netherlands* joining the top three.

Another interesting aspect to highlight is the emergence of bilateral connections. All the arrows are unidirectional in the 2005 scenario (Fig. 4.25). However, in

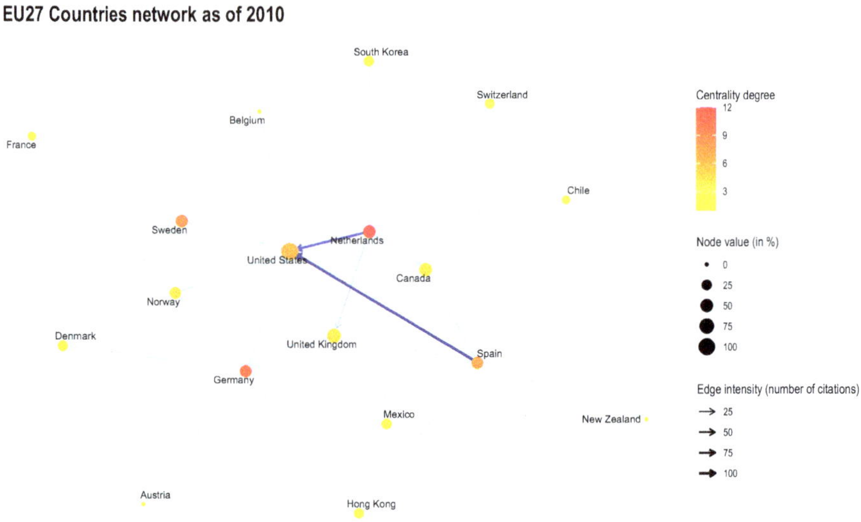


**Fig. 4.26** European countries network analysis as of 2010

**Table 4.27** European countries network analysis as of 2010

| Node | Country | Degree |
|---|---|---|
| *3* | Netherlands | 12 |
| *4* | Germany | 10 |
| *2* | Sweden | 8 |
| *1* | Spain | 7 |
| *7* | United States | 6 |
| *5* | Belgium | 4 |
| *6* | Norway | 4 |
| *8* | United Kingdom | 4 |
| *14* | Canada | 4 |
| *9* | South Korea | 2 |
| *12* | Hong Kong | 2 |
| *16* | Denmark | 2 |
| *18* | Mexico | 2 |
| *10* | New Zealand | 1 |
| *11* | France | 1 |
| *13* | Chile | 1 |
| *15* | Switzerland | 1 |
| *17* | Austria | 1 |

Fig. 4.26, we observe a bidirectional exchange of citations between *Netherlands* and *Spain*.

Figure 4.27 shows even stronger and more extensive connections among European countries, encompassing all papers published up to 2015 by authors affiliated

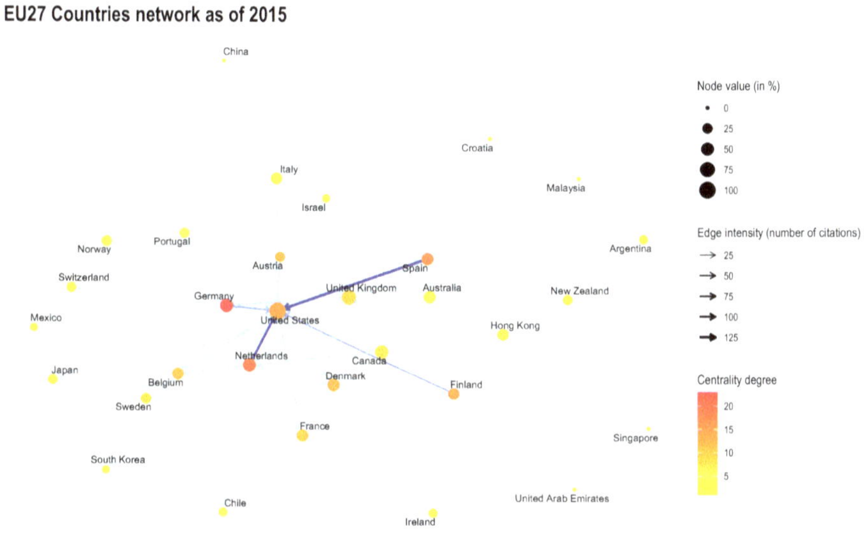


Fig. 4.27 European countries network analysis as of 2015

with European institutions. Thirty-one nations are interconnected by 96 ties, many of which are bidirectional. Notably, new countries such as *Italy* and *Portugal* have joined this network, acting as bridges for new connections both within and outside Europe. For instance, Italy represents the connection to China within the network. The most prevalent connections within this network continue to be directed toward the *United States*, consistent with our previous empirical analysis. The driving countries in terms of centrality are *Germany*, the *Netherlands*, and *Spain*, as summarized in Table 4.28.

The information gathered thus far has provided us with a foundation to delve into the intricate relationships among European affiliations, authors, and journals. The following paragraphs will explore these connections in detail, uncovering the evolved complex network.

### 4.5.3 Network Analysis of European Affiliations

After examining the major countries in the European network of tax avoidance literature, our focus now shifts to understanding the relationships among various affiliations and how they have evolved. Beginning with the first snapshot, encompassing all 160 papers published by authors affiliated with European institutions between the beginning of 2000 and the first quarter of 2023, we present Fig. 4.28. The visualization reveals a tightly knit core, resembling the findings discussed in Sect. 4.4.2 (Fig. 4.6). The most central affiliations, determined by centrality

**Table 4.28** European countries network analysis as of 2015

| Node | Country | Degree |
|---|---|---|
| *9* | Germany | 23 |
| *4* | Netherlands | 18 |
| *1* | Spain | 16 |
| *5* | Finland | 13 |
| *14* | United States | 13 |
| *2* | Denmark | 11 |
| *8* | Austria | 11 |
| *11* | Belgium | 10 |
| *6* | France | 9 |
| *15* | United Kingdom | 8 |
| *7* | Sweden | 7 |
| *26* | Canada | 7 |
| *3* | Italy | 6 |
| *10* | Portugal | 5 |
| *12* | Norway | 4 |
| *25* | Australia | 4 |
| *13* | Switzerland | 3 |
| *20* | New Zealand | 3 |
| *23* | Hong Kong | 3 |
| *27* | Japan | 3 |
| *18* | South Korea | 2 |
| *21* | Israel | 2 |
| *22* | Ireland | 2 |
| *30* | Mexico | 2 |
| *16* | Malaysia | 1 |
| *17* | Argentina | 1 |
| *19* | Singapore | 1 |
| *24* | United Arab Emirates | 1 |
| *28* | Chile | 1 |
| *29* | China | 1 |
| *31* | Croatia | 1 |

degree as detailed in Table 4.29, include *Ludwig-Maximilians-Universität München*, *WHU—Otto Beisheim School of Management*, and *Tilburg University*.

These results align perfectly with our earlier country analysis and, more importantly, provide additional insights that may foreshadow the upcoming network analysis of European authors in the subsequent section. To illustrate this point, let us examine the node *WHU—Otto Beisheim School of Management* in particular. We have conducted an in-depth analysis of each edge to gain a comprehensive understanding of the factors contributing to its centrality. Remarkably, we have discovered that a single author, *Jacob M.*, accounts for 779 out of the institution's 813 connections (95.82%). This observation emphasizes that the centrality of certain institutions is not solely attributed to the institution itself but can heavily depend

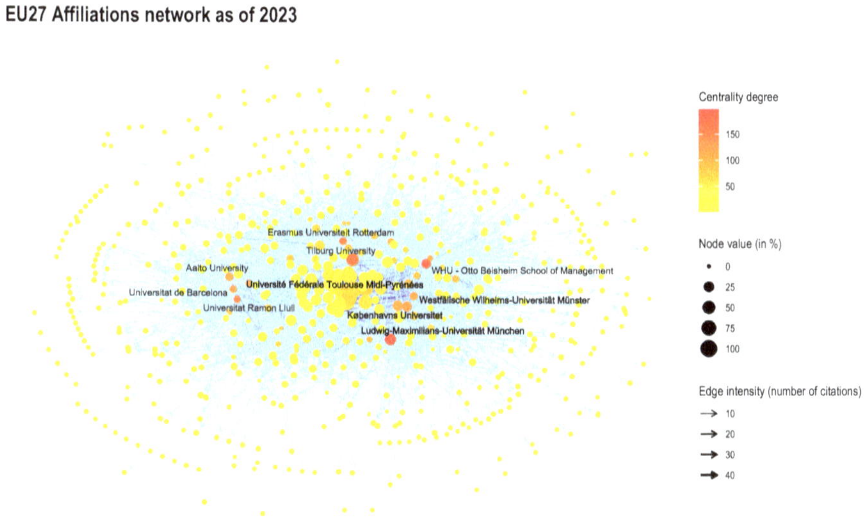


**Fig. 4.28** European affiliations network analysis as of 2023

on the authors affiliated with it. Such insight is crucial to acknowledge, given that affiliations are dynamic variables, as authors may change their affiliations, subsequently impacting the centrality of the respective institution.

To provide a more visually comprehensive representation of the connections among the top affiliations, we have generated Fig. 4.29. This visualization focuses explicitly on the links connecting the top 5% of affiliations based on their centrality degree.

The most central nodes in the network of European affiliations have established a strong connection with *Harvard University* in the United States. This correlation is not surprising, as we have previously discussed the influential role of *Harvard University* in the global network.

#### 4.5.3.1 Time Evolution of European Affiliations Network

We observe a distinct and well-defined situation when examining the earlier period between 2000 and 2005 within the European scope. Figure 4.30 provides a clear overview, revealing the presence of four prominent communities, each revolving around a central node connected to the renowned American university once again. Notably, the central nodes within these communities are primarily European, mainly from Northern European affiliations.

This observation is reinforced by the data presented in Table 4.30, which highlights the primary European affiliations during this period, including *Uppsala Universitet* (Sweden), *Stockholms Universitet* (Sweden), and *Handelshöyskolen BI*

**Table 4.29** European affiliations network analysis as of 2023 (first 30 rows)

| Node | Affiliation | Degree |
|---|---|---|
| *11* | Ludwig-Maximilians-Universität München | 198 |
| *12* | WHU—Otto Beisheim School of Management | 185 |
| *13* | Tilburg University | 160 |
| *16* | Universitat Ramon Llull | 154 |
| *2* | Erasmus Universiteit Rotterdam | 151 |
| *10* | Københavns Universitet | 120 |
| *51* | Université Fédérale Toulouse Midi-Pyrénées | 119 |
| *15* | Universitat de Barcelona | 118 |
| *8* | Westfälische Wilhelms-Universität Münster | 112 |
| *35* | Aalto University | 111 |
| *26* | IE Business School | 107 |
| *44* | Universität Mannheim | 103 |
| *14* | IÉSEG School of Management | 90 |
| *1* | NHH Norwegian School of Economics | 86 |
| *25* | University of Cologne | 84 |
| *29* | Leuphana Universität Lüneburg | 83 |
| *46* | Vrije Universiteit Amsterdam | 82 |
| *4* | Handelshögskolan i Stockholm | 80 |
| *41* | Universität Regensburg | 78 |
| *48* | University of Piraeus | 78 |
| *49* | Université de Montpellier | 78 |
| *42* | Grenoble Ecole de Management | 77 |
| *43* | EM Normandie | 77 |
| *32* | Rijksuniversiteit Groningen | 74 |
| *183* | Harvard University | 73 |
| *5* | Wirtschaftsuniversität Wien | 71 |
| *9* | Goethe-Universität Frankfurt am Main | 71 |
| *23* | Helsingin Yliopisto | 71 |
| *58* | Universiteit van Amsterdam | 71 |
| *45* | Universität Hamburg | 68 |

(Norway). Furthermore, each node relies on the connections within the network, drawing information and knowledge (i.e., citations) from European and international papers.

Upon expanding our analysis to include the papers from the subsequent five-year period, we uncover some notable changes in the landscape. Quantitatively, the number of affiliations involved has increased to 109 (compared to 40 in 2015), and connections have grown from 56 to 196. The communities depicted in Fig. 4.31 are not as distinct as in the earlier period, and the most central affiliations have also shifted, as indicated in Table 4.31.

Taking the lead is now *Universidad Carlos III de Madrid*, followed by *Tilburg University* and *Rijksuniversiteit Groningen*. It is worth noting that two international

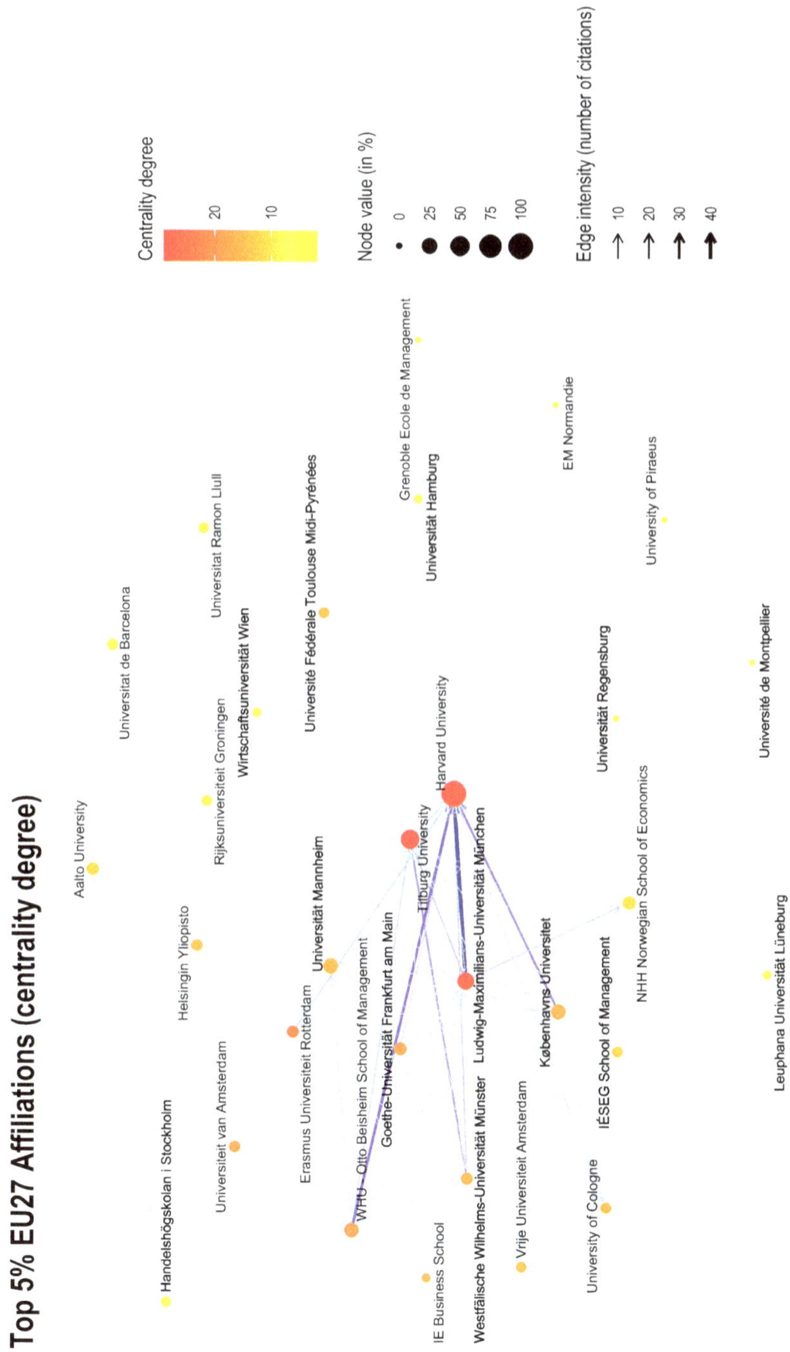


**Fig. 4.29** Top 5% European affiliations network analysis as of 2023 sorted by centrality degree

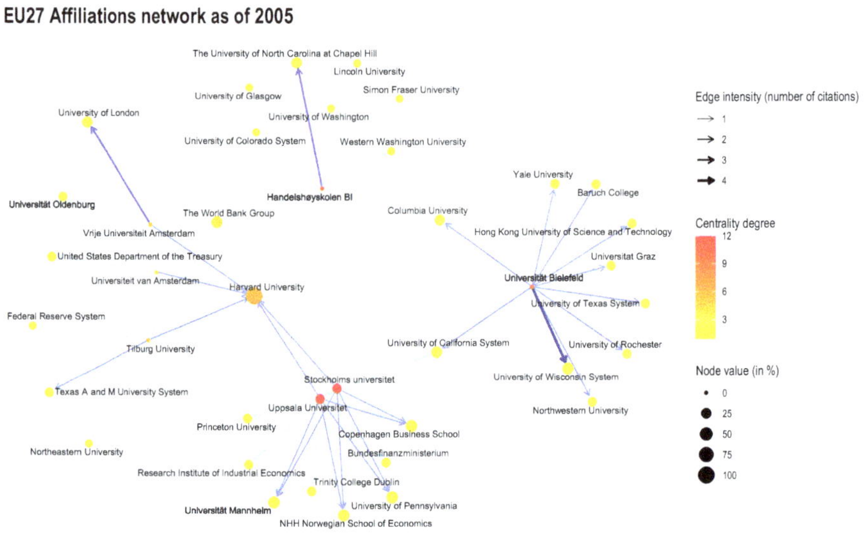


**Fig. 4.30** European affiliations network analysis as of 2005

institutions, namely the *European Commission* with a degree score of 10 and *The World Bank Group* with a degree score of 5, secure the 7th and 13th positions, respectively. Additionally, the *National Bureau of Economic Research* contributes to the network with a degree score of 2, as observed toward the bottom of the table.

Similar to the case of *Jacob M.* discussed earlier, the centrality of this institution is primarily attributed to three authors who, as co-authors of the same paper, have made this affiliation the focal point of the European network during this period. The pivotal paper, titled *The economic determinants of conditional conservatism (SCOPUS-ID: 65149093869)*, was authored by *Lara J.*, *Osma B.*, and *Penalva F.* and was published in 2009 in the *Journal of Business Finance and Accounting*. The paper is important on several dimensions, as it shows that taxation influences both conditional and unconditional conservatism and plays a central role on inter-temporal income shifting. We will revisit these authors in Sect. 4.5.4.1 as they turn out to be the most influential authors in the same period.

The ultimate convergence among the various communities and establishment of a unified network, albeit not yet fully consolidated, is apparent in the third and the final period under analysis, as depicted in Fig. 4.32. From this visualization, distinct groupings are no longer observed, owing to the increased interconnectedness even among previously peripheral nodes that had remained isolated. However, minimal variations are noted among the most influential nodes, with a predominant focus on Spain, specifically *Universidad Carlos III de Madrid* and *Universitat de Barcelona*, following the leading role of *Tilburg University*. The top 30 affiliations are summarized in Table 4.32.

**Table 4.30** European affiliations network analysis as of 2005 (first 30 rows)

| Node | Affiliation | Degree |
|---|---|---|
| *3* | Uppsala Universitet | 12 |
| *4* | Stockholms universitet | 12 |
| *1* | Handelshöyskolen BI | 10 |
| *5* | Universität Bielefeld | 10 |
| *14* | Harvard University | 6 |
| *2* | Vrije Universiteit Amsterdam | 5 |
| *7* | Tilburg University | 5 |
| *6* | Universiteit van Amsterdam | 4 |
| *38* | The World Bank Group | 4 |
| *29* | University of California System | 3 |
| *10* | Princeton University | 2 |
| *12* | Trinity College Dublin | 2 |
| *13* | University of Pennsylvania | 2 |
| *15* | Universität Mannheim | 2 |
| *21* | United States Department of the Treasury | 2 |
| *24* | Copenhagen Business School | 2 |
| *25* | Universität Oldenburg | 2 |
| *30* | NHH Norwegian School of Economics | 2 |
| *33* | Columbia University | 2 |
| *36* | Bundesfinanzministerium | 2 |
| *37* | Research Institute of Industrial Economics | 2 |
| *8* | University of Glasgow | 1 |
| *9* | University of London | 1 |
| *11* | Yale University | 1 |
| *16* | Lincoln University | 1 |
| *17* | Hong Kong University of Science and Technology | 1 |
| *18* | University of Colorado System | 1 |
| *19* | University of Texas System | 1 |
| *20* | University of Washington | 1 |
| *22* | The University of North Carolina at Chapel Hill | 1 |

### 4.5.4 Network Analysis of European Authors

In the European context, the analysis of the authors' network assumes even greater significance. As discussed in the preceding section and indicated in Table 4.33, the limited number of tax avoidance scholars within each affiliation elevates their role as catalysts for the centrality of various universities and institutions.

With an average of 2.36 authors per affiliation and a median of 2, it becomes apparent that individual authors exert substantial influence on their respective affiliations. This holds despite some affiliations having a larger pool of scholars, as

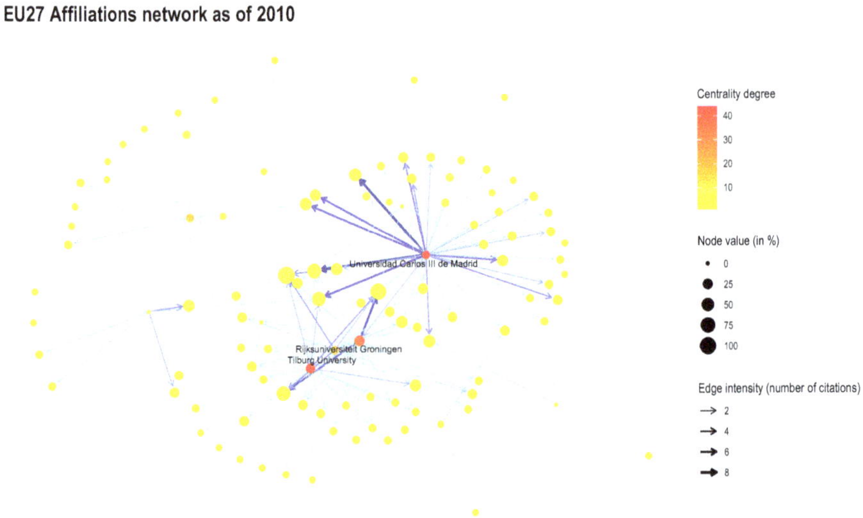


**Fig. 4.31** European affiliations network analysis as of 2010

their centrality stems from the ability to attract connections facilitated by individual actors.

Now that we have established this crucial aspect, we delve into the analysis of the author networks up until the first quarter of 2023. As depicted in Fig. 4.33, the European author network exhibits a discernible core structure, surrounded by nodes that gradually diminish in centrality, expanding the overall sphere. The most central author within the European sphere, as indicated in Table 4.34 and illustrated in the graph, is *Jacob M.*, affiliated with the *WHU—Otto Beisheim School of Management*, boasting a centrality degree of 465.

To better understand the relationships between main contributors, we have conducted a focused analysis of the connections among the top 2% of authors. Figure 4.34 illustrates these links, highlighting noteworthy connections such as the strong relationship between *Jacob M.*, *Hanlon M.*, and *Shevlin T.*, who are also influential authors in the global network. Furthermore, these authors are connected to a select group of individuals, including *Watrin C.*, *De Vito A.*, and *Riedel N.*, among others. These connections play a significant role within the European network.

#### 4.5.4.1 Time Evolution of European Authors Network

Upon further exploration of the temporal evolution of the author network, we have uncovered a pattern resembling the progression of affiliations. This pattern aligns well with what we have discussed before on the greater influence of

**Table 4.31** European affiliations network analysis as of 2010 (first 30 rows)

| Node | Affiliation | Degree |
|---|---|---|
| *1* | Universidad Carlos III de Madrid | 44 |
| *5* | Tilburg University | 43 |
| *4* | Rijksuniversiteit Groningen | 35 |
| *9* | Universiteit van Amsterdam | 17 |
| *2* | Uppsala Universitet | 16 |
| *8* | Universität Bielefeld | 10 |
| *11* | European Commission | 10 |
| *10* | Ludwig-Maximilians-Universität München | 9 |
| *28* | Harvard University | 8 |
| *7* | Stockholms universitet | 7 |
| *33* | University of Texas System | 5 |
| *69* | University of California System | 5 |
| *99* | The World Bank Group | 5 |
| *3* | Göteborgs Universitet | 4 |
| *6* | Vrije Universiteit Amsterdam | 4 |
| *12* | Handelshöyskolen BI | 4 |
| *44* | University of Toronto | 4 |
| *46* | The University of North Carolina at Chapel Hill | 4 |
| *50* | University of Wisconsin System | 4 |
| *57* | Texas A and M University System | 4 |
| *70* | NHH Norwegian School of Economics | 4 |
| *72* | The Regents of the University of Michigan | 4 |
| *84* | The University of Chicago | 4 |
| *108* | Northwestern University | 4 |
| *17* | Pennsylvania State University | 3 |
| *22* | Arizona State University | 3 |
| *27* | University of Pennsylvania | 3 |
| *43* | United States Department of the Treasury | 3 |
| *55* | National Bureau of Economic Research | 3 |
| *63* | University of Connecticut | 3 |

individual authors within the European context, which is why the prominence of their respective institutions.

Figure 4.35 vividly illustrates the presence of five distinct communities arranged in a radial configuration around a central node. By 2005, 73 interconnected authors were identified, linked together through 118 edges. Following our analysis of affiliations, the most central nodes (as illustrated in Table 4.35) consist of European authors who predominantly establish connections with their American counterparts. These American colleagues are significant collaboration partners for the European authors.

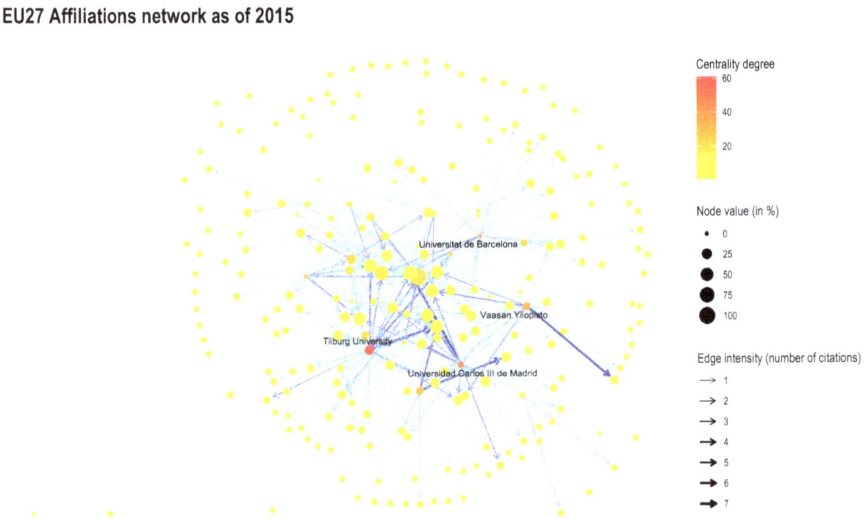


**Fig. 4.32** European affiliations network analysis as of 2015

Including papers published between 2006 and 2010 in the analysis of network dynamics has revealed a new scenario, which appears to be more consolidated, as evident from the right-hand side of Fig. 4.36.

During this period, driven by increased productivity and heightened interest in tax avoidance, we observe a rise in the number of nodes (authors) predominantly within the extensive sphere of connections bridging the Old Continent and researchers affiliated with major American institutions (see Table 4.36).

As we examine the cluster structure across all the papers up to 2015, it becomes apparent that it has transformed into a cohesive network. Figure 4.37 presents a central nucleus encompassing all the authors, gradually solidifying over time. During this period, the focal point of European research is between Northern Europe, the Netherlands, and Spain, as evident from Table 4.37.

During this period, the literature's progression has contributed to Europe's increasing centrality on the global stage. By 2023, a substantial network of connections has emerged, providing a platform for new researchers and scholars to exchange knowledge and explore various aspects of the intricate phenomenon of tax avoidance.

While momentarily setting aside the examination of journal networks, which we have already explored in Sect. 4.4.4 and provided valuable insights into the central themes within the tax avoidance literature, we can conclude these three studies with the broader analysis discussed earlier. Specifically, within the European scope, the literature is predominantly shaped by individual authors rather than by interconnections between nations or affiliations.

**Table 4.32** European affiliations network analysis as of 2015 (first 30 rows)

| Node | Affiliation | Degree |
|---|---|---|
| *4* | Tilburg University | 61 |
| *8* | Universidad Carlos III de Madrid | 44 |
| *5* | Universitat de Barcelona | 35 |
| *6* | Vaasan Yliopisto | 31 |
| *15* | Goethe-Universität Frankfurt am Main | 27 |
| *22* | Rijksuniversiteit Groningen | 25 |
| *12* | Aalto University | 23 |
| *1* | Universitat Ramon Llull | 22 |
| *19* | Ludwig-Maximilians-Universität München | 22 |
| *67* | Harvard University | 20 |
| *13* | Universiteit van Amsterdam | 19 |
| *16* | Universidad de la Laguna | 19 |
| *28* | Universität Mannheim | 18 |
| *11* | Helsingin Yliopisto | 17 |
| *107* | University of California System | 17 |
| *7* | Université Paris 1 Panthéon-Sorbonne | 16 |
| *14* | Universität Innsbruck | 16 |
| *2* | Københavns Universitet | 15 |
| *17* | Technische Universität Dortmund | 14 |
| *9* | Uppsala Universitet | 13 |
| *20* | Administrative Headquarters of the Max Planck Society | 13 |
| *29* | European Commission | 13 |
| *23* | Otto von Guericke University of Magdeburg | 12 |
| *26* | Universität Bielefeld | 12 |
| *106* | Massachusetts Institute of Technology | 12 |
| *10* | Osterreichisches Institut fur Wirtschaftsforschung | 11 |
| *21* | Vrije Universiteit Amsterdam | 11 |
| *27* | Universität Hamburg | 11 |
| *51* | University of London | 10 |
| *121* | University of Toronto | 10 |

This highlights a crucial lesson for European researchers who wish to delve into this subject. Instead of relying on broader affiliations or national boundaries, it is essential to prioritize the search for individual European authors. They serve as invaluable sources of knowledge and connections, offering a more focused and fruitful approach to study development. By emphasizing the expertise and insights of these authors, researchers can unlock new perspectives and make significant advancements in the tax field.

**Table 4.33** Number of authors for each European Affiliation as of 2023 (first 50 rows)

| Author parent affiliation | Number of authors |
|---|---|
| Ludwig-Maximilians-Universität München | 18 |
| Universität Mannheim | 10 |
| Tilburg University | 8 |
| Westfälische Wilhelms-Universität Münster | 8 |
| Écoles polytechniques fédérales | 6 |
| Erasmus Universiteit Rotterdam | 6 |
| Københavns Universitet | 6 |
| Universitat Ramon Llull | 6 |
| Uppsala Universitet | 6 |
| Vrije Universiteit Amsterdam | 6 |
| Handelshögskolan i Stockholm | 5 |
| NHH Norwegian School of Economics | 5 |
| European Commission | 4 |
| Goethe-Universität Frankfurt am Main | 4 |
| Università degli Studi di Padova | 4 |
| Wirtschaftsuniversität Wien | 4 |
| Aalto University | 3 |
| Administrative Headquarters of the Max Planck Society | 3 |
| Eberhard Karls Universität Tübingen | 3 |
| Freie Universität Berlin | 3 |
| Friedrich-Alexander-Universität Erlangen-Nürnberg | 3 |
| Helsingin Yliopisto | 3 |
| Nyenrode Business University | 3 |
| Rijksuniversiteit Groningen | 3 |
| Universidad Carlos III de Madrid | 3 |
| Universidade do Minho | 3 |
| Università degli Studi di Napoli Federico II | 3 |
| Universität Bielefeld | 3 |
| Universitat de Barcelona | 3 |
| Universiteit van Amsterdam | 3 |
| University of Cologne | 3 |
| Deutsche Bundesbank | 2 |
| EDHEC Business School | 2 |
| Gottfried Wilhelm Leibniz Universität Hannover | 2 |
| Hanken School of Economics | 2 |
| Humboldt-Universität zu Berlin | 2 |
| IÉSEG School of Management | 2 |
| Katholische Universität Eichstätt—Ingolstadt | 2 |
| KU Leuven | 2 |
| Leuphana Universität Lüneburg | 2 |
| Otto von Guericke University of Magdeburg | 2 |

(continued)

**Table 4.33** (continued)

| Author parent affiliation | Number of authors |
|---|---|
| SciencesPo Paris | 2 |
| Technische Universität Berlin | 2 |
| Universidad de Murcia | 2 |
| Università Bocconi | 2 |
| Universitat Graz | 2 |
| Universität Hamburg | 2 |
| Universität Innsbruck | 2 |
| Universität Kassel | 2 |
| Universität Paderborn | 2 |

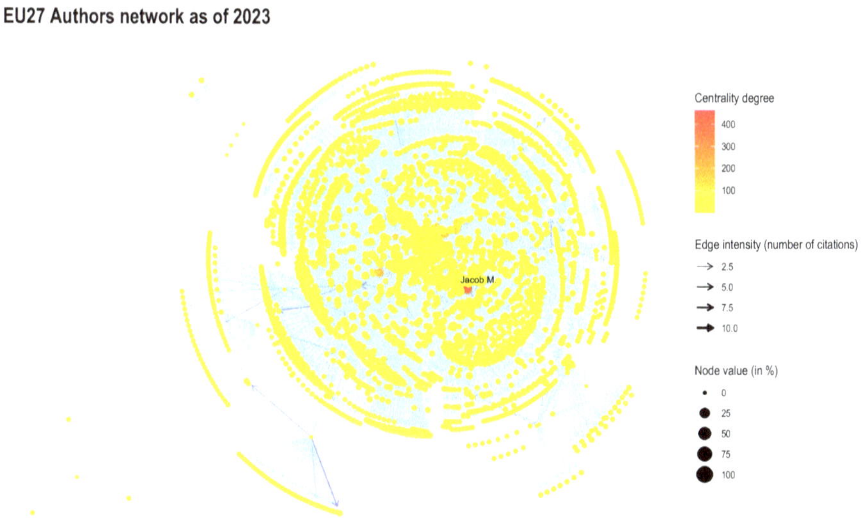


**Fig. 4.33** European authors network analysis as of 2023

## *4.5.5 Network Analysis of European Journals*

The analysis of journals based on papers from the European perimeter reveals results consistent with the global findings discussed in Sect. 4.4.4. There are no significant variations in terms of the centrality of the topics discussed in Europe compared to the rest of the world. However, when examining Fig. 4.38, we observe a slight heterogeneity in the connections between the journals. Notably, there are distinct and often bilateral links among the most central journals, as indicated in Table 4.38.

**Table 4.34** European affiliations network analysis as of 2023 (first 20 rows)

| Node | Author | Degree |
|---|---|---|
| *18* | Jacob M. | 465 |
| *53* | Hardeck I. | 217 |
| *21* | Ravenda D. | 211 |
| *22* | Argilés-Bosch J.M. | 211 |
| *36* | De Vito A. | 193 |
| *14* | Watrin C. | 191 |
| *13* | Thomsen M. | 165 |
| *77* | García-Bland6n J. | 161 |
| *90* | Schanz D. | 151 |
| *41* | Kovermann J. | 147 |
| *42* | Velte P. | 147 |
| *62* | Benlemlih M. | 139 |
| *60* | Peillex J. | 138 |
| *61* | Jaballah J. | 138 |
| *72* | Delis M.D. | 138 |
| *83* | Law K.K.F. | 138 |
| *20* | Ruiz C.V. | 136 |
| *52* | Sundvik D. | 124 |
| *106* | Huizinga H. | 124 |
| *23* | Valencia-Silva M.M. | 123 |

The image depicts multiple connections between the journals *European Accounting Review*, *Contemporary Accounting Research*, *Accounting and Business Research*, *Journal of Public Economics*, and *Journal of International Accounting, Auditing and Taxation*. These nodes possess two noteworthy characteristics that warrant scrutiny. First, all these journals primarily focus on accounting, albeit with distinct nuances. Second, as indicated in the table, none of these journals holds a 4* ranking according to the ABS classification. This further proves that the ranking does not solely determine a journal's centrality.

#### 4.5.5.1 Time Evolution of European Journals

The temporal progression of the European journals network has followed a similar pattern to the global network, although with a focus on the initial development of distinct communities and a gradual convergence over the years.

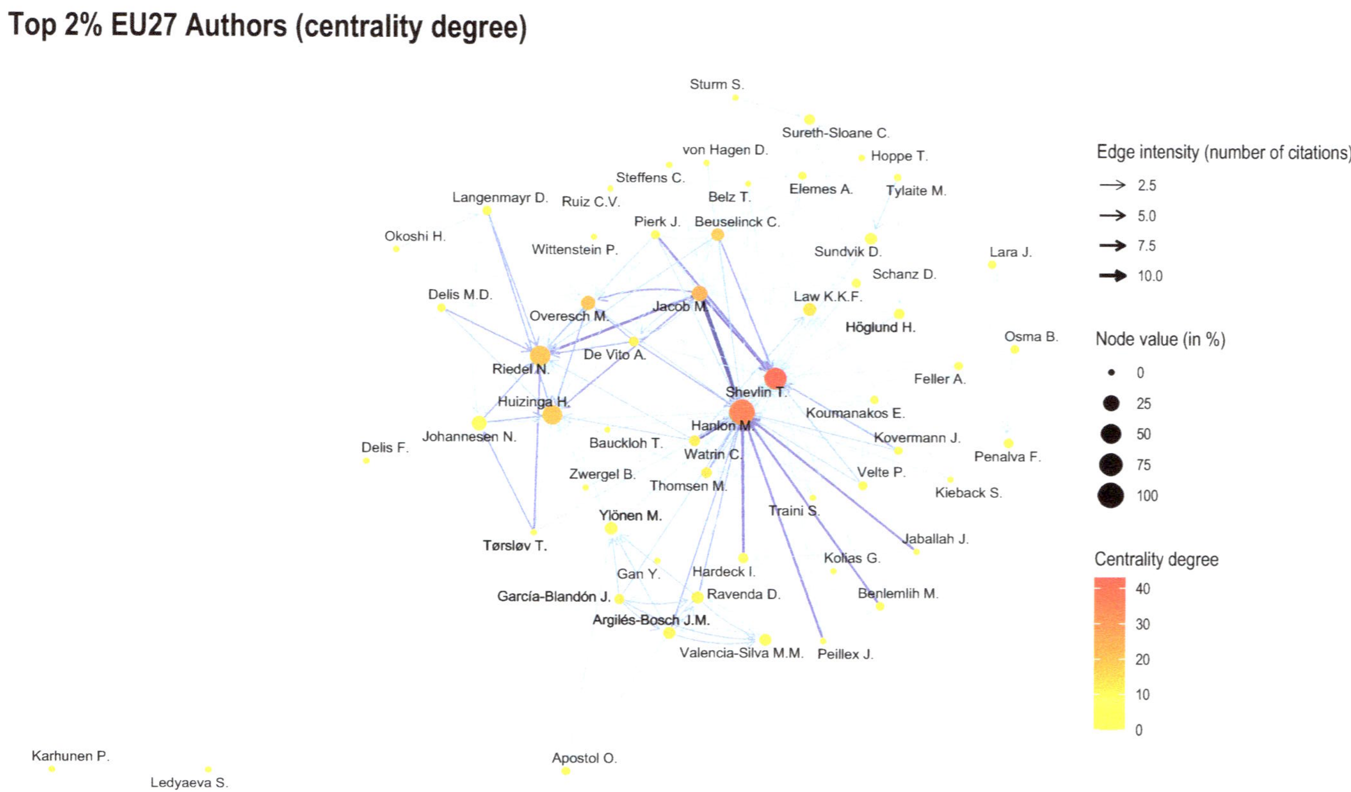


**Fig. 4.34** Top 2% European authors network analysis as of 2023 sorted by centrality degree

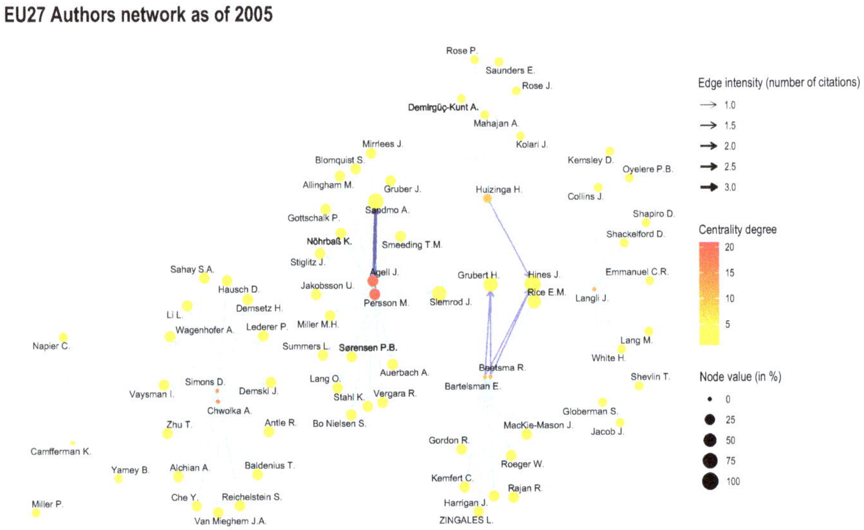


**Fig. 4.35** European authors network analysis as of 2005

**Table 4.35** European affiliations network analysis as of 2005 (first 20 rows)

| Node | Author | Degree |
|---|---|---|
| *3* | Agell J. | 21 |
| *4* | Persson M. | 21 |
| *5* | Chwolka A. | 15 |
| *6* | Simons D. | 15 |
| *1* | Langli J. | 13 |
| *7* | Beetsma R. | 11 |
| *8* | Bartelsman E. | 11 |
| *9* | Huizinga H. | 10 |
| *69* | Slemrod J. | 5 |
| *20* | Hines J. | 4 |
| *52* | Rice E.M. | 4 |
| *2* | Camfferman K. | 3 |
| *30* | Grubert H. | 3 |
| *14* | Stiglitz J. | 2 |
| *15* | Blomquist S. | 2 |
| *16* | Li L. | 2 |
| *17* | Mirrlees J. | 2 |
| *18* | Allingham M. | 2 |
| *19* | Sandmo A. | 2 |
| *21* | Lang O. | 2 |

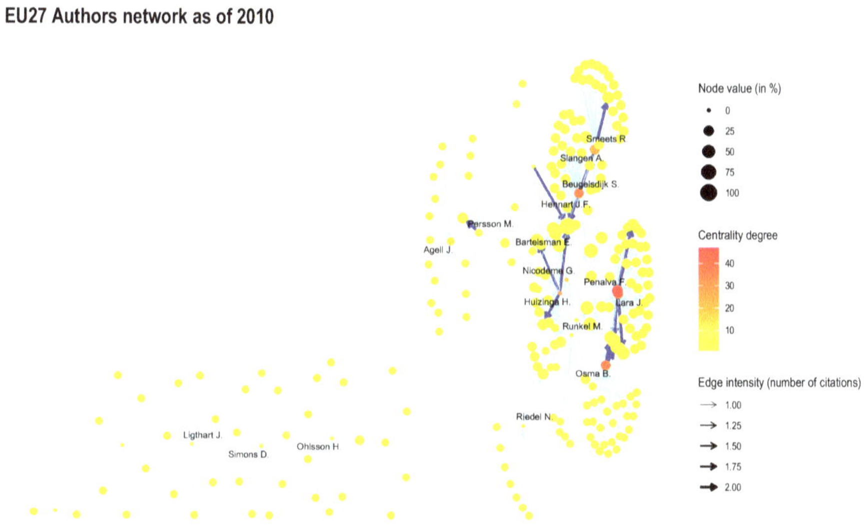


Fig. 4.36 European authors network analysis as of 2010

**Table 4.36** European affiliations network analysis as of 2010 (first 20 rows)

| Node | Author | Degree |
|---|---|---|
| 2 | Penalva F. | 47 |
| 1 | Lara J. | 46 |
| 7 | Osma B. | 37 |
| 5 | Beugelsdijk S. | 36 |
| 6 | Hennart J.F. | 34 |
| 14 | Huizinga H. | 29 |
| 15 | Slangen A. | 26 |
| 16 | Smeets R. | 24 |
| 18 | Nicodeme G. | 18 |
| 11 | Agell J. | 13 |
| 19 | Riedel N. | 11 |
| 3 | Ohlsson H. | 9 |
| 9 | Persson M. | 9 |
| 13 | Simons D. | 9 |
| 21 | Bartelsman E. | 8 |
| 17 | Runkel M. | 7 |
| 22 | Ligthart J. | 7 |
| 10 | Chwolka A. | 6 |
| 43 | Hines J. | 6 |
| 57 | Shleifer A. | 6 |

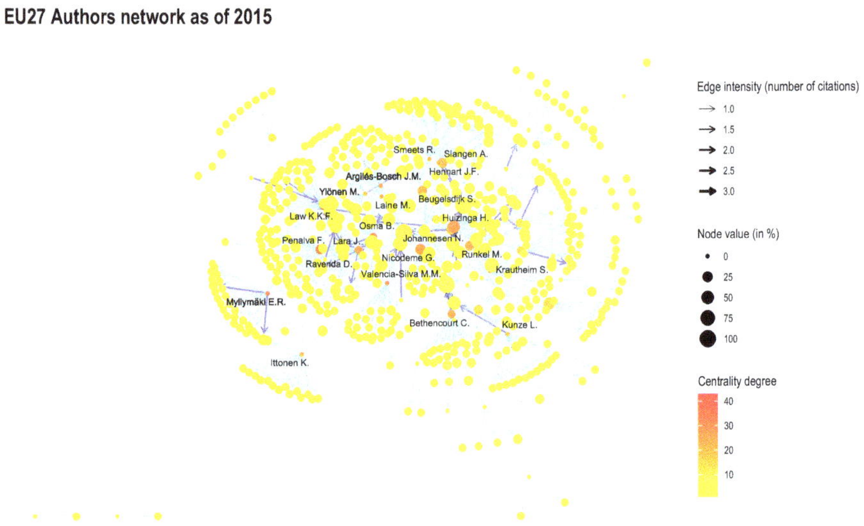


**Fig. 4.37** European authors network analysis as of 2015

**Table 4.37** European authors network analysis as of 2015 (first 20 rows)

| Node | Author | Degree |
|---|---|---|
| *24* | Law K.K.F. | 43 |
| *10* | Penalva F. | 35 |
| *7* | Myllymäki E.R. | 34 |
| *5* | Ravenda D. | 28 |
| *12* | Osma B. | 28 |
| *1* | Valencia-Silva M.M. | 27 |
| *6* | Argilés-Bosch J.M. | 27 |
| *8* | Huizinga H. | 26 |
| *20* | Lara J. | 25 |
| *42* | Nicodeme G. | 25 |
| *15* | Laine M. | 24 |
| *21* | Bethencourt C. | 23 |
| *32* | Runkel M. | 22 |
| *31* | Beugelsdijk S. | 21 |
| *16* | Slangen A. | 20 |
| *17* | Hennart J.F. | 20 |
| *19* | Krautheim S. | 20 |
| *39* | Smeets R. | 20 |
| *2* | Johannesen N. | 18 |
| *14* | Ylönen M. | 18 |

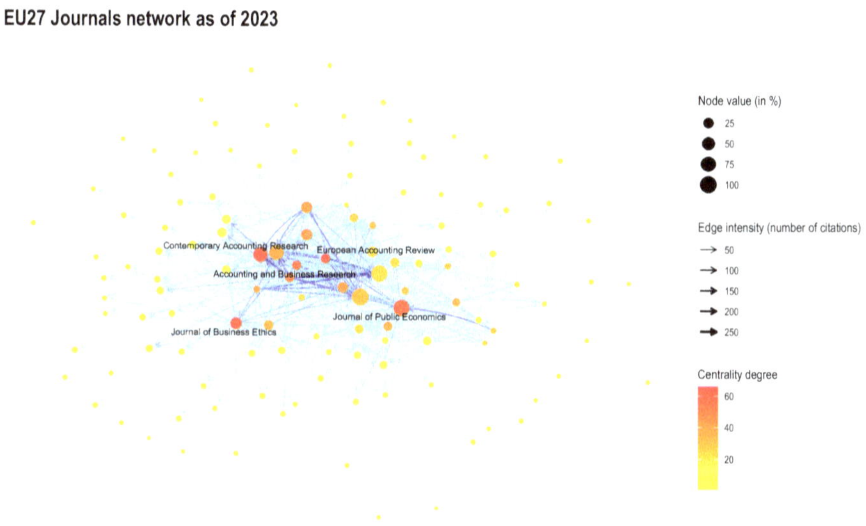


**Fig. 4.38** European journals network analysis as of 2023

**Table 4.38** European journals network analysis as of 2023 (first 20 rows)

| Node | Journal | ABS | Degree |
|---|---|---|---|
| *4* | European Accounting Review | 3 | 66 |
| *3* | Contemporary Accounting Research | 4 | 62 |
| *10* | Journal of Business Ethics | 3 | 61 |
| *12* | Accounting and Business Research | 3 | 58 |
| *7* | Journal of Public Economics | 3 | 57 |
| *11* | Journal of International Accounting, Auditing and Taxation | 3 | 53 |
| *2* | Journal of International Business Studies | 4* | 41 |
| *6* | Review of Accounting Studies | 4 | 40 |
| *9* | Accounting Review | 4* | 39 |
| *5* | Journal of Business Finance and Accounting | 3 | 36 |
| *19* | Journal of Accounting Research | 4* | 36 |
| *27* | Quarterly Journal of Economics | 4* | 34 |
| *8* | American Economic Journal: Economic Policy | 3 | 33 |
| *14* | Critical Perspectives on Accounting | 3 | 31 |
| *22* | Journal of Financial and Quantitative Analysis | 4 | 29 |
| *33* | Journal of Financial Economics | 4* | 28 |
| *15* | Canadian Journal of Economics | 3 | 26 |
| *23* | Review of Economics and Statistics | 4 | 26 |
| *17* | Scandinavian Journal of Economics | 3 | 25 |
| *24* | Review of Financial Studies | 4* | 25 |

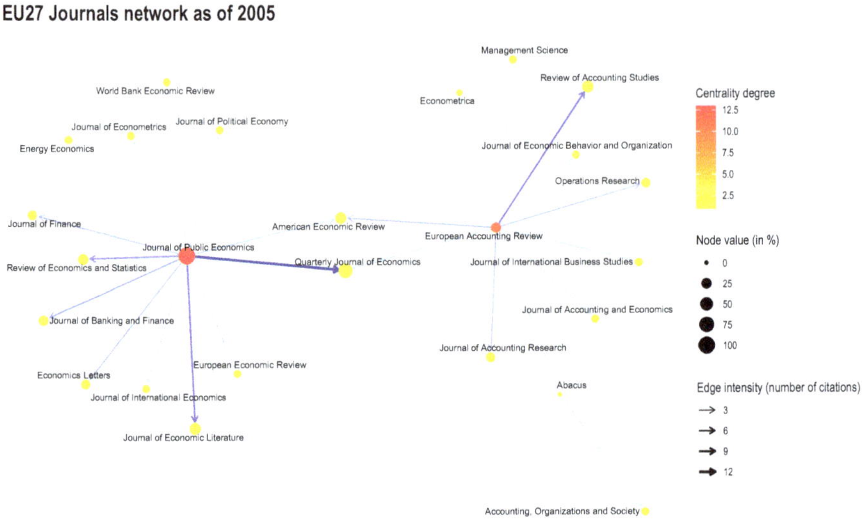


**Fig. 4.39** European journals network analysis as of 2005

During the first 5 years, the network comprised 25 journals and 25 links, consisting of tax avoidance papers authored by individuals in the European Area and their corresponding references. Figure 4.39 illustrates a sparse network characterized by predominantly one-sided links, with two central journals, namely, the *Journal of Public Economics* and the *European Accounting Review*. This observation is further supported by the summary in Table 4.39, where most considered journals exhibit a centrality degree of 1.

As we dig deeper into the study of tax avoidance determinants and potential measures from 2006 through 2010, the coverage of our paper corpus expands slightly to encompass an additional 13 journals. Furthermore, the surge in scholarly output during this time frame has increased the number of connections between the various publishers, reaching a total of 52.

**Table 4.39** European journals network analysis as of 2005

| Node | Journal | ABS | Degree |
|---|---|---|---|
| *3* | Journal of Public Economics | 3 | 13 |
| *1* | European Accounting Review | 3 | 11 |
| *2* | Abacus | 3 | 2 |
| *7* | American Economic Review | 4* | 2 |
| *9* | Quarterly Journal of Economics | 4* | 2 |
| *4* | Accounting, Organizations and Society | 4* | 1 |
| *5* | Journal of Econometrics | 4 | 1 |
| *6* | Operations Research | 4* | 1 |
| *8* | Journal of Economic Behavior and Organization | 3 | 1 |
| *10* | Journal of Accounting Research | 4* | 1 |
| *11* | Review of Accounting Studies | 4 | 1 |
| *12* | Journal of Accounting and Economics | 4* | 1 |
| *13* | Journal of International Business Studies | 4* | 1 |
| *14* | Energy Economics | 3 | 1 |
| *15* | Journal of Banking and Finance | 3 | 1 |
| *16* | Journal of Economic Literature | 4 | 1 |
| *17* | European Economic Review | 3 | 1 |
| *18* | Economics Letters | 3 | 1 |
| *19* | Management Science | 4* | 1 |
| *20* | Journal of International Economics | 4 | 1 |
| *21* | World Bank Economic Review | 3 | 1 |
| *22* | Journal of Finance | 4* | 1 |
| *23* | Review of Economics and Statistics | 4 | 1 |
| *24* | Journal of Political Economy | 4* | 1 |
| *25* | Econometrica | 4* | 1 |

This development is visually represented in Fig. 4.40, which portrays a progressive convergence and closeness among the nodes while still displaying clusters. From an analytical standpoint, Table 4.40 reaffirms this finding, exhibiting higher centrality degree scores than the previous period. Although there have been shifts in the rankings for several journals, the *Journal of Public Economics* maintains its position as the predominant journal.

Reflecting on the year 2015, we can observe the initial upsurge in tax avoidance literature within Europe, which has continued to gain momentum in subsequent years up to the present. The communities of journals have become increasingly interconnected and supportive, aligning with the recognition of tax avoidance as a multidisciplinary phenomenon that intersects with various domains such as economics, finance, and accounting (Hanlon & Heitzman, 2010).

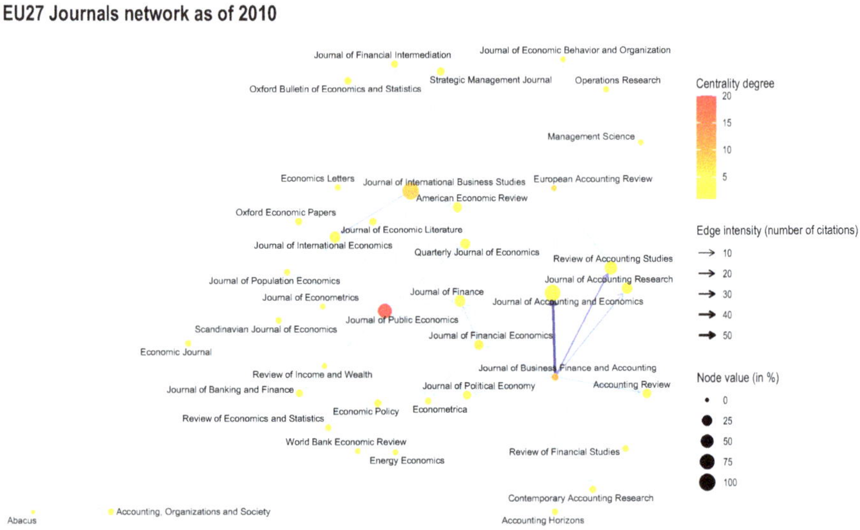


**Fig. 4.40** European journals network analysis as of 2010

Notably, in Fig. 4.41, we come across journals like *Management Science* or *Journal of Human Resources*, which were previously unassociated with tax avoidance research. However, as Table 4.41 reveals, the centrality of the network remains unchanged and focused on the *Journal of Public Economics*. This is primarily due to the numerous publications this journal has received over time in tax avoidance.

## 4.6 Summary

This chapter has focused on the empirical analysis of tax avoidance research networks. It has covered the sample selection process and the utilization of the Scopus API for data retrieval. The chapter has offered descriptive statistics, gathering insights into research productivity by country, institutions, journals listed in the ABS journal list, and individual authors.

The chapter has highlighted the significance of network analysis techniques in examining the interconnections and collaborations among countries, affiliations, authors, and journals within the tax avoidance research domain. By applying the network analysis, we have uncovered hidden patterns, identified key players and clusters, and developed a deeper understanding of the overall tax research landscape.

**Table 4.40** European journals network analysis as of 2010 (first 25 rows)

| Node | Journal | ABS | Degree |
|---|---|---|---|
| *2* | Journal of Public Economics | 3 | 20 |
| *1* | Journal of Business Finance and Accounting | 3 | 10 |
| *3* | Journal of International Business Studies | 4* | 8 |
| *5* | European Accounting Review | 3 | 8 |
| *6* | Journal of Financial Economics | 4* | 6 |
| *10* | Journal of Finance | 4* | 4 |
| *22* | Quarterly Journal of Economics | 4* | 4 |
| *7* | Scandinavian Journal of Economics | 3 | 3 |
| *11* | Journal of International Economics | 4 | 3 |
| *12* | Journal of Accounting and Economics | 4* | 3 |
| *20* | American Economic Review | 4* | 3 |
| *8* | Journal of Accounting Research | 4* | 2 |
| *13* | Review of Accounting Studies | 4 | 2 |
| *29* | Journal of Political Economy | 4* | 2 |
| *39* | Econometrica | 4* | 2 |
| *4* | Abacus | 3 | 1 |
| *9* | Journal of Population Economics | 3 | 1 |
| *14* | Accounting, Organizations and Society | 4* | 1 |
| *15* | Journal of Econometrics | 4 | 1 |
| *16* | Review of Income and Wealth | 3 | 1 |
| *17* | Operations Research | 4* | 1 |
| *18* | Contemporary Accounting Research | 4 | 1 |
| *19* | Accounting Review | 4* | 1 |
| *21* | Journal of Economic Behavior and Organization | 3 | 1 |
| *23* | Energy Economics | 3 | 1 |

Moreover, the chapter has presented a focused analysis of the European network, showcasing the application of data cleaning and alignment techniques. By studying the network of European countries, affiliations, authors, and journals, we have gained valuable insights into the dynamics specific to this region, shedding light on collaborations and knowledge dissemination within the European network.

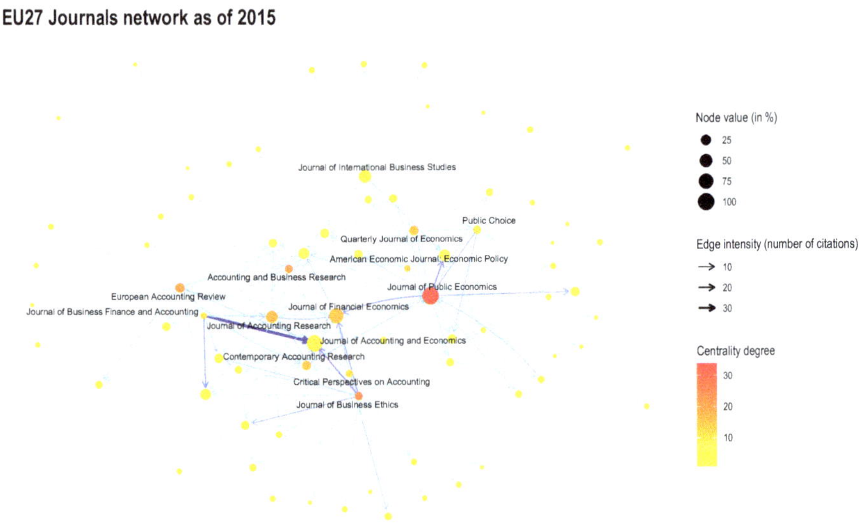


**Fig. 4.41** European journals network analysis as of 2015

**Table 4.41** European journals network analysis as of 2015 (first 25 rows)

| Node | Journal | ABS | Degree |
|---|---|---|---|
| *3* | Journal of Public Economics | 3 | 34 |
| *1* | Journal of Business Ethics | 3 | 24 |
| *4* | Accounting and Business Research | 3 | 21 |
| *6* | European Accounting Review | 3 | 19 |
| *11* | Journal of Accounting Research | 4* | 17 |
| *15* | Quarterly Journal of Economics | 4* | 15 |
| *16* | Journal of Financial Economics | 4* | 15 |
| *5* | Contemporary Accounting Research | 4 | 14 |
| *2* | American Economic Journal: Economic Policy | 3 | 13 |
| *7* | Journal of Business Finance and Accounting | 3 | 11 |
| *8* | Critical Perspectives on Accounting | 3 | 11 |
| *9* | Journal of International Business Studies | 4* | 9 |
| *10* | Public Choice | 3 | 9 |

(continued)

**Table 4.41** (continued)

| Node | Journal | ABS | Degree |
|---|---|---|---|
| *30* | Journal of Accounting and Economics | 4* | 9 |
| *14* | Journal of International Accounting, Auditing and Taxation | 3 | 8 |
| *17* | Journal of Comparative Economics | 3 | 8 |
| *24* | Journal of Finance | 4* | 7 |
| *28* | Journal of International Economics | 4 | 6 |
| *33* | Review of Financial Studies | 4* | 6 |
| *36* | American Economic Review | 4* | 6 |
| *37* | Econometrica | 4* | 6 |
| *19* | Review of Economics and Statistics | 4 | 5 |
| *27* | Review of Accounting Studies | 4 | 5 |
| *34* | Journal of Economic Literature | 4 | 5 |
| *18* | Canadian Journal of Economics | 3 | 4 |

## References


Bajo, E., Barbi, M., & Hillier, D. (2020). Where should i publish to get promoted? A finance journal ranking based on business school promotions. *Journal of Banking & Finance, 114*, 105780.

Cohen, L., Frazzini, A., & Malloy, C. (2008). The small world of investing: Board connections and mutual fund returns. *Journal of Political Economy, 116*(5), 951–979.

Desai, M. A., & Dharmapala, D. (2006). Corporate tax avoidance and high-powered incentives. *Journal of Financial Economics, 79*(1), 145–179.

Desai, M. A., & Dharmapala, D. (2009). Corporate tax avoidance and firm value. *Review of Economics and Statistics, 91*(3), 537–546.

Dowling, G. R. (2014). The curious case of corporate tax avoidance: Is it socially irresponsible? *Journal of Business Ethics, 124*, 173–184.

Fox, Z. D., Jacob, M., Wilde, J. H., & Wilson, R. J. (2022). Beyond borders: Uncertainty in supragovernmental tax enforcement and corporate investment. *The Accounting Review, 97*(6), 233–261.

Goyal, S., Van Der Leij, M. J., & Moraga-González, J. L. (2006). Economics: An emerging small world. *Journal of Political Economy, 114*(2), 403–412.

Hanlon, M., & Heitzman, S. (2010). A review of tax research. *Journal of Accounting and Economics, 50*(2), 127–178.

Jacob, M. (2022). Real effects of corporate taxation: A review. *European Accounting Review, 31*(1), 269–296.

Kong, X., Shi, Y., Yu, S., Liu, J., & Xia, F. (2019). Academic social networks: Modeling, analysis, mining and applications. *Journal of Network and Computer Applications, 132*, 86–103.

Muschelli, J. (2019). rscopus: Scopus database 'API' interface. R package version 0.6, 6.

Scholes, M. S., Wolfson, M. A., Erickson, M., Maydew, E., & Shevlin, T. (2014). *Taxes & business strategy*. Prentice Hall.

Shackelford, D. A., & Shevlin, T. (2001). Empirical tax research in accounting. *Journal of Accounting and Economics, 31*(1–3), 321–387.

Wilde, J. H., & Wilson, R. J. (2018). Perspectives on corporate tax planning: Observations from the past decade. *Journal of the American Taxation Association, 40*(2), 63–81.

# Chapter 5
# Conclusion

As we conclude our comprehensive exploration of the tax avoidance research landscape, we are equipped with valuable insights and knowledge that can serve as a compass for young academics venturing into this multidisciplinary field. Throughout this monograph, we have examined the network dynamics of authors, affiliations, and journals, delving into temporal evolutions and interconnections. In this concluding section, we will synthesize our findings, highlight key takeaways, and provide actionable strategies for aspiring researchers to navigate the complex research area of tax avoidance.

First, it is central to understand the role of individual authors. One crucial piece of evidence from our analysis is the significant influence that individual authors can have, particularly in the European tax avoidance research scene. We have witnessed how specific scholars, rather than affiliations or countries, drive the centrality and connectivity of the associated network. This insight emphasizes how important it is to focus on individual European authors as critical sources of knowledge and connections in this field. Aspiring academics can benefit from identifying and engaging with influential authors to foster collaborations, access valuable expertise, and expand their research networks.

Second, the tax avoidance research area has a dynamic setting regarding affiliations. Indeed, our examination has revealed an interesting pattern where small groups of scholars within institutions contribute to the overall centrality of universities and research organizations. As a direct consequence of the first point above, this finding also underscores the need for young academics to recognize individual scholars' potential influence and expertise within their respective affiliations. Aspiring academics can tap into a wealth of knowledge and establish themselves as contributors within the tax avoidance research community by actively seeking out and collaborating with these influential researchers.

Third, the identification of leading journals is also critical. Analyzing journal networks has allowed us to identify trends and interrelationships among tax avoidance publications. While the centrality of topics discussed in European journals



A. De Vito, F. Grossetti, *Tax Avoidance Research*, SIDREA Series in Accounting and Business Administration, https://doi.org/10.1007/978-3-031-51765-5_5

aligns with the global landscape, we have noticed nuanced differences in the links between various publications. Aspiring academics must stay abreast of these network dynamics and familiarize themselves with key journals that dominate the field. Engaging with these journals can enhance their visibility, credibility, and potential for impactful contributions.

Fourth, as the influential reviews of Shackelford and Shevlin (2001), Hanlon and Heitzman (2010), and recently Wilde and Wilson (2018) have well emphasized, tax avoidance research is interdisciplinary. This field transcends traditional disciplinary boundaries, intersecting with economics, finance, accounting, and law. Our findings confirm the multidisciplinary nature of this field, highlighting how adopting an interdisciplinary approach is key to success when conducting research. Aspiring academics are encouraged to broaden their horizons, collaborate across disciplines, fields, and methods, and draw insights from a diversified set of perspectives. Embracing interdisciplinarity will enrich their understanding of tax avoidance and allow a more comprehensive exploration of this complex phenomenon.

Fifth, leveraging the power of technology and information systems in such a fast-paced world has become essential nowadays. Our analysis has relied on network visualizations and data-driven techniques to gain deeper insights into the tax avoidance research landscape. Technology and quantitative methods have become invaluable for researchers, enabling them to analyze vast amounts of data, identify complex patterns, and uncover hidden connections. Emerging academics should embrace such advancements, develop data analysis skills and machine learning techniques, and leverage available tools to enhance their research methodologies and discover new avenues for investigation.

Sixth, tax avoidance research can and must emphasize the relevance of policy implications. As tax avoidance significantly impacts society, policymakers, and businesses, aspiring academics should strive to bridge the gap between academic research, policy, and practical application. By examining these implications, understanding the regulatory landscape, and engaging with stakeholders, researchers can contribute to evidence-based policy discussions and offer valuable insights to address the challenges of tax avoidance. This applied focus will enhance the social relevance of their work and foster collaborations with policymakers and industry experts.

In conclusion, our exploration of the tax avoidance research landscape has provided valuable insights and pathways for aspiring academics embarking on their journey in this field. By recognizing the centrality of individual authors, understanding the dynamics of affiliations and journals, embracing interdisciplinarity, leveraging technology, and emphasizing policy implications, aspiring researchers can navigate the complex tax avoidance research landscape effectively. The knowledge gained from this comprehensive analysis serves as a roadmap to guide future researchers toward impactful contributions. As they embark on their academic pursuits, we encourage emerging academics to seize the opportunities presented by this vibrant and evolving field, driving positive change and shaping the discourse on tax avoidance.

Looking ahead, several promising avenues for future research can further enhance our understanding of tax avoidance and its implications. By addressing these research gaps, one can contribute to the ongoing discourse and drive meaningful advancements in this complex field. For instance, as tax laws and regulations evolve, new strategies for tax avoidance may emerge. Future research should identify and analyze these emerging strategies, examining their effectiveness, implications, and potential countermeasures. This research can help policymakers and tax authorities stay ahead of evolving tax avoidance practices and develop proactive measures to mitigate their impact.

Technology, including digital platforms, distributed systems such as blockchain, cryptocurrencies, and other forms of digital assets (Gietzmann & Grossetti, 2021), has transformed the business landscape and introduced new challenges and opportunities for tax avoidance (Cong et al., 2023). Future research can investigate the intersection of technology and tax avoidance, exploring how advancements like blockchain, artificial intelligence, and data analytics affect tax planning, enforcement, and regulatory measures. Understanding the implications of technology-driven tax avoidance will be crucial for policymakers and tax authorities to design effective counter-strategies.

In addition, tax avoidance has wide-ranging societal and economic implications. Future research should quantify and evaluate the economic costs of tax avoidance, especially its more aggressive forms, on government revenue, public services, income inequality, and economic development (Alstadsæter et al., 2019, 2022). Additionally, studying the social consequences, such as the erosion of public trust and the impact on social welfare programs, can provide valuable insights for policymakers and stakeholders in addressing these challenges.

Tax avoidance practices vary across jurisdictions due to differences in tax systems, regulations, and enforcement mechanisms. Future research can adopt a comparative and international approach, analyzing tax avoidance trends, strategies, and policy responses in different countries and regions, as the extant empirical tax literature is primarily dominated by US-centered research (Bruehne & Jacob, 2019). This cross-jurisdictional analysis can highlight best practices, regulatory gaps, and opportunities for international cooperation to curb tax avoidance effectively. Closely related future research can also evaluate the effectiveness of anti-avoidance measures, including legislative initiatives, international agreements, and enforcement practices (Alexander et al., 2020). Analyzing the impact of anti-avoidance measures on taxpayer behavior, revenue collection, and the overall tax system will help policymakers refine existing measures and develop targeted approaches to combat tax avoidance more effectively.

By addressing these potential avenues for future research, academics can not only expand the knowledge base on tax avoidance but also inform policy debates and contribute to developing effective strategies for addressing this pervasive issue. Continued interdisciplinary collaboration, data-driven as well as quantitative analyses, and stakeholder engagement will be crucial in advancing research and fostering positive change.

## References

Alexander, A., De Vito, A., & Jacob, M. (2020). Corporate tax reforms and tax-motivated profit shifting: evidence from the eu. *Accounting and Business Research, 50*(4), 309–341.

Alstadsæter, A., Johannesen, N., Herry, S. L. G., & Zucman, G. (2022). Tax evasion and tax avoidance. *Journal of Public Economics, 206*, 104587.

Alstadsæter, A., Johannesen, N., & Zucman, G. (2019). Tax evasion and inequality. *American Economic Review, 109*(6), 2073–2103.

Bruehne, A., & Jacob, M. (2019). Corporate tax avoidance and the real effects of taxation: A review. *TRR 266 Accounting for Transparency Working Paper Series No. 34.*

Cong, L. W., Landsman, W., Maydew, E., & Rabetti, D. (2023). Tax-loss harvesting with cryptocurrencies. *Journal of Accounting and Economics* 101607.

Gietzmann, M., & Grossetti, F. (2021). Blockchain and other distributed ledger technologies: Where is the accounting? *Journal of Accounting and Public Policy, 40*(5), 106881.

Hanlon, M., & Heitzman, S. (2010). A review of tax research. *Journal of Accounting and Economics, 50*(2), 127–178.

Shackelford, D. A., & Shevlin, T. (2001). Empirical tax research in accounting. *Journal of Accounting and Economics, 31(1–3)*, 321–387.

Wilde, J. H., & Wilson, R. J. (2018). Perspectives on corporate tax planning: Observations from the past decade. *Journal of the American Taxation Association, 40*(2), 63–81.

# Appendix A
# Sample of Papers

See Table A.1.

**Table A.1** The full list of papers used for the analyses

| N | Title | Authors | Journal | Year |
|---|---|---|---|---|
| 1 | A Further Examination Of Income Shifting Through Transfer Pricing Considering Firm Size And/Or Distress | Conover T.L., Nichols N.B. | International Journal of Accounting | 2000 |
| 2 | Tax Arbitrage And Labor Supply | Agell J., Persson M. | Journal of Public Economics | 2000 |
| 3 | The Influence Of International Status And Operating Segments On Firms' Choice Of Bonus Plans | Carnes G., Guffey D. | Journal of International Accounting, Auditing and Taxation | 2000 |
| 4 | Collusion Proof Transfer Payment Schemes With Multiple Agents | Balachandran K.R., Li S.H. | Review of Quantitative Finance and Accounting | 2000 |
| 5 | The Cost Of Equality: Unequal Bequests And Tax Avoidance | McGarry K. | Journal of Public Economics | 2001 |
| 6 | Transfer Pricing Of Intangible Property Harmony And Discord Across Five Countries | Borkowski S.C. | International Journal of Accounting | 2001 |
| 7 | The Taxation Of Domestic And Foreign Banking | Demirgüç–Kunt A., Huizinga H. | Journal of Public Economics | 2001 |
| 8 | The Effect Of Marginal Tax Rates On Taxable Income: A Panel Study Of The 1988 Tax Flattening In Canada | Sillamaa M., Veall M. | Journal of Public Economics | 2001 |

(continued)



A. De Vito, F. Grossetti, *Tax Avoidance Research*, SIDREA Series in Accounting and Business Administration, https://doi.org/10.1007/978-3-031-51765-5

**Table A.1** (continued)

| N | Title | Authors | Journal | Year |
|---|---|---|---|---|
| 9 | Tax Evasion And Self-Insurance | Lee K. | Journal of Public Economics | 2001 |
| 10 | Redistribution When Avoidance Behavior Is Heterogeneous | Kopczuk W. | Journal of Public Economics | 2001 |
| 11 | Searching For Ghosts: Who Are The Nonfilers And How Much Tax Do They Owe? | Erard B., Ho C. | Journal of Public Economics | 2001 |
| 12 | Tax And Incentive Trade-Offs In Multinational Transfer Pricing | Smith M. | Journal of Accounting, Auditing & Finance | 2002 |
| 13 | The Optimal Elasticity Of Taxable Income | Kopczuk W., Slemrod J. | Journal of Public Economics | 2002 |
| 14 | Differences In The Reported Performances Of Foreign-Controlled And Domestically-Controlled Firms: Some U.K. Evidence | Emmanuel C.R., Oyelere P.B. | Journal of International Accounting, Auditing and Taxation | 2002 |
| 15 | International Income-Shifting Regulations: Empirical Evidence From Australia And Canada | Cloyd C.B. | International Journal of Accounting | 2003 |
| 16 | Firm Valuation Effects Of The Expatriation Of U.S. Corporations To Tax-Haven Countries | Weiss I. | Journal of the American Taxation Association | 2003 |
| 17 | International Income-Shifting Regulations: Empirical Evidence From Australia And Canada | Eldenburg L., Pickering J., Yu W.W. | International Journal of Accounting | 2003 |
| 18 | Firm Valuation Effects Of The Expatriation Of U.S. Corporations To Tax-Haven Countries | Cloyd C., Mills L., Weaver C. | Journal of the American Taxation Association | 2003 |
| 19 | The Transformation Of An Accounting System During Organisational Change: A Longitudinal Study | Gurd B., Thorne H. | Financial Accountability and Management | 2003 |
| 20 | Generalised R-Based And S-Based Taxes Under Uncertainty | Bond S.R., Devereux M.P. | Journal of Public Economics | 2003 |
| 21 | Impacts Of Revenue Sharing, Profit Sharing And Transfer Pricing On Quality-Improving Investments | Chwolka A., Simons D. | European Accounting Review | 2003 |
| 22 | Tax-Motivated Transfer Pricing And Us Intrafirm Trade Prices | Clausing K. | Journal of Public Economics | 2003 |
| 23 | Why Pay More? Corporate Tax Avoidance Through Transfer Pricing In Oecd Countries | Bartelsman E., Beetsma R. | Journal of Public Economics | 2003 |
| 24 | Are Individual Investors Tax Savvy? Evidence From Retail And Discount Brokerage Accounts | Barber B., Odean T. | Journal of Public Economics | 2004 |

(continued)

**Table A.1** (continued)

| N | Title | Authors | Journal | Year |
|---|---|---|---|---|
| 25 | The Profits Of The Dutch East India Company'S Japan Trade | Camfferman K., Cooke T. | Abacus | 2004 |
| 26 | On The Timeliness Of Tax Reform | Hines J. | Journal of Public Economics | 2004 |
| 27 | Income Shifting, Investment, And Tax Competition: Theory And Evidence From Provincial Taxation In Canada | Mintz J., Smart M. | Journal of Public Economics | 2004 |
| 28 | The Impact Of Separate Taxation On The Intra-Household Allocation Of Assets: Evidence From The Uk | Stephens M., Ward-Batts J. | Journal of Public Economics | 2004 |
| 29 | The Costs Of Shared Ownership: Evidence From International Joint Ventures | Desai M.A., Foley C.F., Hines J.R. | Journal of Financial Economics | 2004 |
| 30 | Taxable Income Differences Between Foreign And Domestic Controlled Corporations In Norway | Langli J., Saudagaran S. | European Accounting Review | 2004 |
| 31 | Fiscal Incentives, Corporate Structure And Financial Aspects Of Treasury Management Operations | Stewart J. | Accounting Forum | 2005 |
| 32 | South Pacific Tax Havens: From Leaders In The Race To The Bottom To Laggards In The Race To The Top? | Sharman J. | Accounting Forum | 2005 |
| 33 | Mobile People, Mobile Capital And Tax Neutrality: Sustaining A Market For Offshore Finance Centres | Rawlings G. | Accounting Forum | 2005 |
| 34 | Top Incomes In The United States And Canada Over The Twentieth Century | Saez E. | Journal of the European Economic Association | 2005 |
| 35 | Comparing The Distribution Of Top Incomes Across Countries | Atkinson A.B. | Journal of the European Economic Association | 2005 |
| 36 | The Role Of Accountancy Firms In Tax Avoidance: Some Evidence And Issues | Hampton M.P., Sikka P. | Accounting Forum | 2005 |
| 37 | Estimating The Magnitude Of Capital Flight Due To Abnormal Pricing In International Trade: The Russia-Usa Case | De Boyrie M.E., Pak S.J., Zdanowicz J.S. | Accounting Forum | 2005 |
| 38 | Corporate Tax Avoidance And High-Powered Incentives | Desai M., Dharmapala D. | Journal of Financial Economics | 2006 |
| 39 | The Demand For Tax Haven Operations | Desai M., Foley C., Hines J. | Journal of Public Economics | 2006 |
| 40 | External And Internal Pricing In Multidivisional Firms | Baldenius T., Reichelstein S. | Journal of Accounting Research | 2006 |

(continued)

**Table A.1** (continued)

| N | Title | Authors | Journal | Year |
|---|---|---|---|---|
| 41 | Tax Shelters And Corporate Debt Policy | Graham J.R., Tucker A.L. | Journal of Financial Economics | 2006 |
| 42 | Tax Avoidance And Intra-Family Transfers | Nordblom K., Ohlsson H. | Journal of Public Economics | 2006 |
| 43 | The Corporation As A Tax Shelter: Evidence From Recent Israeli Tax Changes | Romanov D. | Journal of Public Economics | 2006 |
| 44 | Where'S The Harm In Tax Competition?. Lessons From Us Multinationals In Ireland | Killian S. | Critical Perspectives on Accounting | 2006 |
| 45 | Tax Misreporting And Avoidance By Nonprofit Organizations | Omer T., Yetman R. | Journal of the American Taxation Association | 2007 |
| 46 | Company Tax Reform With A Water'S Edge | Riedel N., Runkel M. | Journal of Public Economics | 2007 |
| 47 | Revenue Sharing And Information Exchange Under Non-Discriminatory Taxation | Keen M., Ligthart J. | Scandinavian Journal of Economics | 2007 |
| 48 | Bequest And Tax Planning: Evidence From Estate Tax Returns | Kopczuk W. | Quarterly Journal of Economics | 2007 |
| 49 | The History Of Pata And Its Effect On Advance Pricing Arrangements And Mutual Agreement Procedures | Borkowski S. | Journal of International Accounting, Auditing and Taxation | 2008 |
| 50 | Perspectives On China'S Outward Foreign Direct Investment | Morck R., Yeung B., Zhao M. | Journal of International Business Studies | 2008 |
| 51 | Capital Structure And International Debt Shifting | Huizinga H., Laeven L., Nicodeme G. | Journal of Financial Economics | 2008 |
| 52 | International Profit Shifting Within Multinationals: A Multi-Country Perspective | Huizinga H., Laeven L. | Journal of Public Economics | 2008 |
| 53 | Do Countries Compete Over Corporate Tax Rates? | Devereux M., Lockwood B., Redoano M. | Journal of Public Economics | 2008 |
| 54 | Earnings Management Through Effective Tax Rates: The Effects Of Tax-Planning Investment And The Sarbanes-Oxley Act Of 2002 | Cook K., Huston G., Omer T. | Contemporary Accounting Research | 2008 |
| 55 | International Financial Institutions And The New Global Managerial Order | Murphy J. | Critical Perspectives on Accounting | 2008 |
| 56 | Financial Flows And Treasury Management Firms In Ireland | Stewart J. | Accounting Forum | 2008 |

(continued)

**Table A.1** (continued)

| N | Title | Authors | Journal | Year |
|---|---|---|---|---|
| 57 | Rules And Accounting: Vagueness In Conceptual Frameworks | Penno M.C. | Accounting Horizons | 2008 |
| 58 | Tax Law Changes, Income-Shifting And Measured Wage Inequality: Evidence From India | Sivadasan J., Slemrod J. | Journal of Public Economics | 2008 |
| 59 | Friction In Related-Party Trade When A Rival Is Also A Customer | Arya A., Mittendorf B., Yoon D. | Management Science | 2008 |
| 60 | Internal Pricing | Baldenius T. | Foundations and Trends in Accounting | 2008 |
| 61 | Taxable Income As A Performance Measure: The Effects Of Tax Planning And Earnings Quality | Ayers B.C., Jiang J., Laplante S.K. | Contemporary Accounting Research | 2009 |
| 62 | What Does Tax Aggressiveness Signal? Evidence From Stock Price Reactions To News About Tax Shelter Involvement | Hanlon M., Slemrod J. | Journal of Public Economics | 2009 |
| 63 | Discussion Of "Taxable Income As A Performance Measure: The Effects Of Tax Planning And Earnings Quality" | Raedy J.S. | Contemporary Accounting Research | 2009 |
| 64 | Competition And Corporate Tax Avoidance: Evidence From Chinese Industrial Firms | Cai H., Liu Q. | Economic Journal | 2009 |
| 65 | The Economic Determinants Of Conditional Conservatism | Lara J., Osma B., Penalva F. | Journal of Business Finance and Accounting | 2009 |
| 66 | Linking Ethics And Risk Management In Taxation: Evidence From An Exploratory Study In Ireland And The Uk | Doyle E.M., Glaister K.W., Hughes J.F. | Journal of Business Ethics | 2009 |
| 67 | The Response Of Firms To Eligibility Thresholds: Evidence From The Japanese Value-Added Tax | Onji K. | Journal of Public Economics | 2009 |
| 68 | Optimal Scheduling And Incentive Compatible Pricing For A Service System With Quality Of Service Guarantees | Hsu V., Jukic B., Xu S. | Manufacturing and Service Operations Management | 2009 |
| 69 | Tax Motivated Income Shifting And Korean Business Groups (Chaebol) | Jung K., Kim B., Kim B. | Journal of Business Finance and Accounting | 2009 |
| 70 | Corporate Tax Avoidance And Firm Value | Desai M.A., Dharmapala D. | Review of Economics and Statistics | 2009 |
| 71 | Tax Sensitivity And Home State Preferences In Internet Purchasing | Ellison G., Ellison S. | American Economic Journal: Economic Policy | 2009 |

(continued)

**Table A.1** (continued)

| N | Title | Authors | Journal | Year |
|---|---|---|---|---|
| 72 | Which Countries Become Tax Havens? | Dharmapala D., Hines J.R. | Journal of Public Economics | 2009 |
| 73 | Fairness In International Trade And Investment: North American Perspectives | Bird F., Vance T., Woolstencroft P. | Journal of Business Ethics | 2009 |
| 74 | Using Financial Accounting Data To Examine The Effect Of Foreign Operations Located In Tax Havens And Other Countries On U.S. Multinational Firms' Tax Rates | Dyreng S., Lindsey B. | Journal of Accounting Research | 2009 |
| 75 | Tax Competition With Parasitic Tax Havens | Slemrod J., Wilson J.D. | Journal of Public Economics | 2009 |
| 76 | Decentralization, Transfer Pricing, And Tacit Collusion | Chen H., Shor M. | Contemporary Accounting Research | 2009 |
| 77 | Transfer Pricing Practices Of Transnational Corporations In Pata Countries | Borkowski S. | Journal of International Accounting, Auditing and Taxation | 2010 |
| 78 | The Micro-Geography Of Tax Avoidance: Evidence From Littered Cigarette Packs In Chicago | Merriman D. | American Economic Journal: Economic Policy | 2010 |
| 79 | Two Tax Policies Toward Tax Havens: Effects On Production And Capital Location Decisions | Halperin R., Srinidhi B. | Journal of Accounting, Auditing and Finance | 2010 |
| 80 | Are Family Firms More Tax Aggressive Than Non-Family Firms? | Chen S., Chen X., Cheng Q., Shevlin T. | Journal of Financial Economics | 2010 |
| 81 | The Role Of Transfer-Pricing Schemes In Coordinated Supply Chains | Balachandran K.R., Shu-Hsing L.I., Wang H.W., Wang T. | Journal of Accounting, Auditing and Finance | 2010 |
| 82 | International Corporate Taxation And U.S. Multinationals' Behaviour: An Integrated Approach | Azémar C. | Canadian Journal of Economics | 2010 |
| 83 | Voluntary Disclosure Of Auditor-Provided Tax Service Fees | Bedard J., Falsetta D., Krishnamoorthy G., Omer T. | Journal of the American Taxation Association | 2010 |
| 84 | The Dark Side Of Transfer Pricing: Its Role In Tax Avoidance And Wealth Retentiveness | Sikka P., Willmott H. | Critical Perspectives on Accounting | 2010 |
| 85 | Measuring The Location Of Production In A World Of Intangible Productive Assets, Fdi, And Intrafirm Trade | Lipsey R. | Review of Income and Wealth | 2010 |

(continued)

**Table A.1** (continued)

| N | Title | Authors | Journal | Year |
|---|---|---|---|---|
| 86 | Credit Ratings And Taxes: The Effect Of Book-Tax Differences On Ratings Changes | Ayers B., Laplante S., McGuire S. | Contemporary Accounting Research | 2010 |
| 87 | Corporate Tax Harmonization In The Eu | Bettendorf L., de Mooij R., Devereux M., Loretz S., van der Horst A. | Economic Policy | 2010 |
| 88 | Smoke And Mirrors: Corporate Social Responsibility And Tax Avoidance | Sikka P. | Accounting Forum | 2010 |
| 89 | Is There A December Effect? Strategic Prepayments Of Deductible State Income Tax | Shon J., Veliotis S. | Journal of the American Taxation Association | 2010 |
| 90 | Tax, Financial Reporting, And Tunneling Incentives For Income Shifting: An Empirical Analysis Of The Transfer Pricing Behavior Of Chinese-Listed Companies | Firth M., Lo A., Wong R. | Journal of the American Taxation Association | 2010 |
| 91 | Humpty Dumpty'S Guide To Tax Law: Rules, Principles And Certainty In Taxation | James M. | Critical Perspectives on Accounting | 2010 |
| 92 | Value-Added Taxes, Chain Effects, And Informality | de Paula Á., Scheinkman J. | American Economic Journal: Macroeconomics | 2010 |
| 93 | Customer-Driven Vs. Retailer-Driven Search: Channel Performance And Implications | Anupindi R., Li J. | Manufacturing and Service Operations Management | 2010 |
| 94 | Why And How Fdi Stocks Are A Biased Measure Of Mne Affiliate Activity | Beugelsdijk S., Hennart J.F., Slangen A., Smeets R. | Journal of International Business Studies | 2010 |
| 95 | The Market For Corporate Tax Knowledge | Hasseldine J., Holland K., van der Rijt P. | Critical Perspectives on Accounting | 2011 |
| 96 | Reported Trade Figure Discrepancy, Regulatory Arbitrage, And Round-Tripping: Evidence From The China-Hong Kong Trade Data | Fung H.G., Yau J., Zhang G. | Journal of International Business Studies | 2011 |
| 97 | Heterogeneous Firms, 'Profit Shifting' Fdi And International Tax Competition | Krautheim S., Schmidt-Eisenlohr T. | Journal of Public Economics | 2011 |
| 98 | Profit Shifting And Measured Productivity Of Multinational Firms | Maffini G., Mokkas S. | Oxford Bulletin of Economics and Statistics | 2011 |

(continued)

**Table A.1** (continued)

| N | Title | Authors | Journal | Year |
|---|---|---|---|---|
| 99 | The Role Of Multinational Companies In Tax Evasion And Tax Avoidance: The Case Of Nigeria | Otusanya O. | Critical Perspectives on Accounting | 2011 |
| 100 | Income Shifting Within A Dual Income Tax System: Evidence From The Finnish Tax Reform Of 1993 | Pirttilä J., Selin H. | Scandinavian Journal of Economics | 2011 |
| 101 | The Influence Of International Taxation Structures On Corporate Financial Disclosure Patterns | Taylor G., Tower G., Van Der Zahn M. | Accounting Forum | 2011 |
| 102 | The Informal Sector: An Equilibrium Model And Some Empirical Evidence From Brazil | De Paula A., Scheinkman J.A. | Review of Income and Wealth | 2011 |
| 103 | Corporate Tax Avoidance And Stock Price Crash Risk: Firm-Level Analysis | Kim J.B., Li Y., Zhang L. | Journal of Financial Economics | 2011 |
| 104 | Earnings Management Induced By Tax Planning: The Case Of Portuguese Private Firms | Craig R., Marques M., Rodrigues L.L. | Journal of International Accounting, Auditing and Taxation | 2011 |
| 105 | Corporate Taxes And The Location Of Intangible Assets Within Multinational Firms | Dischinger M., Riedel N. | Journal of Public Economics | 2011 |
| 106 | Corporate Income Taxation Of Multinationals In A General Equilibrium Model | Eichner T., Runkel M. | Journal of Public Economics | 2011 |
| 107 | Strategic Transfer Pricing And Social Welfare Under Product Differentiation | Matsui K. | European Accounting Review | 2011 |
| 108 | Is There An Association Between Earnings Management And Auditor-Provided Tax Services? | Krishnan G., Visvanathan G. | Journal of the American Taxation Association | 2011 |
| 109 | The Effects Of Increased Book-Tax Difference Tax Return Disclosures On Firm Valuation And Behavior | Donohoe M., McGill G. | Journal of the American Taxation Association | 2011 |
| 110 | Short-Term Incentive Effects Of A Reduction In The Nol Carryback Period | Albring S.M., Dhaliwal D.S., Khurana I.K., Pereira R. | Journal of the American Taxation Association | 2011 |
| 111 | Foreign (In)Direct Investment And Corporate Taxation | Wamser G. | Canadian Journal of Economics | 2011 |
| 112 | Cost Allocation In Manufacturing-Remanufacturing Operations | Toktay L.B., Wei D. | Production and Operations Management | 2011 |
| 113 | Majority Voting And The Welfare Implications Of Tax Avoidance | Traxler C. | Journal of Public Economics | 2012 |

(continued)

**Table A.1** (continued)

| N | Title | Authors | Journal | Year |
|---|---|---|---|---|
| 114 | Judicial Interpretation Of The Will Of The State: A Hegelian Perspective In The Context Of Taxation | Norton S.D. | Critical Perspectives on Accounting | 2012 |
| 115 | Fin 48, Uncertainty And Transfer Pricing: (Im)Perfect Together? | Borkowski S., Gaffney M. | Journal of International Accounting, Auditing and Taxation | 2012 |
| 116 | Domestic Income Shifting By Chinese Listed Firms | Shevlin T., Tang T.Y.H., Wilson R.J. | Journal of the American Taxation Association | 2012 |
| 117 | Optimal Fiscal Barriers To International Economic Integration In The Presence Of Tax Havens | Johannesen N. | Journal of Public Economics | 2012 |
| 118 | Equity Risk Incentives And Corporate Tax Aggressiveness | Rego S.O., Wilson R. | Journal of Accounting Research | 2012 |
| 119 | Tax Planning, Corporate Governance And Equity Value | Abdul Wahab N., Holland K. | British Accounting Review | 2012 |
| 120 | Tax Professionals' Responsibility For Fraud Detection: The Effects Of Engagement Type And Audit Status | Harrison P.D., Schnee E.J., Todd Dezoort F. | Accounting Horizons | 2012 |
| 121 | Responsibility In Paradise? The Adoption Of Csr Tools By Companies Domiciled In Tax Havens | Preuss L. | Journal of Business Ethics | 2012 |
| 122 | The Effect Of Foreign Reinvestment And Financial Reporting Incentives On Cross-Jurisdictional Income Shifting | Klassen K., Laplante S. | Contemporary Accounting Research | 2012 |
| 123 | Religious Social Norms And Corporate Financial Reporting | Dyreng S., Mayew W., Williams C. | Journal of Business Finance and Accounting | 2012 |
| 124 | Tax Avoidance And Underleverage Puzzle: Korean Evidence | Lim Y. | Review of Quantitative Finance and Accounting | 2012 |
| 125 | Consistency Of Book-Tax Differences And The Information Content Of Earnings | Chen L.H., Dhaliwal D.S., Trombley M.A. | Journal of the American Taxation Association | 2012 |
| 126 | Profit Shifting And Trade Agreements In Imperfectly Competitive Markets | Bagwell K., Staiger R.W. | International Economic Review | 2012 |
| 127 | The Heterogeneous Geographic And Socioeconomic Incidence Of Cigarette Taxes: Evidence From Nielsen Homescan Data | Harding M., Leibtag E., Lovenheim M.F. | American Economic Journal: Economic Policy | 2012 |

(continued)

**Table A.1** (continued)

| N | Title | Authors | Journal | Year |
|---|---|---|---|---|
| 128 | Are U.S. Multinational Corporations Becoming More Aggressive Income Shifters? | Klassen K., Laplante S. | Journal of Accounting Research | 2012 |
| 129 | Competition And Cost Accounting | Narayanan V.G., Smith M. | Foundations and Trends in Accounting | 2012 |
| 130 | Transfer Pricing Or Formula Apportionment? Tax-Induced Distortions Of Multinationals' Investment And Production Decisionś | Martini J.T., Niemann R., Simons D. | Contemporary Accounting Research | 2012 |
| 131 | The Impact Of Thin-Capitalization Rules On The Capital Structure Of Multinational Firms | Buettner T., Overesch M., Schreiber U., Wamser G. | Journal of Public Economics | 2012 |
| 132 | Earnings Shocks And Tax-Motivated Income-Shifting: Evidence From European Multinationals | Dharmapala D., Riedel N. | Journal of Public Economics | 2013 |
| 133 | Institutional Shareholders' Investment Horizons And Tax Avoidance | Khurana I.K., Moser W.J. | Journal of the American Taxation Association | 2013 |
| 134 | The Determinants Of Thinly Capitalized Tax Avoidance Structures: Evidence From Australian Firms | Richardson G., Taylor G. | Journal of International Accounting, Auditing and Taxation | 2013 |
| 135 | Smoke And Mirrors: Corporate Social Responsibility And Tax Avoidance-A Reply To Hasseldine And Morris | Sikka P. | Accounting Forum | 2013 |
| 136 | Corporate Social Responsibility And Tax Avoidance: A Comment And Reflection | Hasseldine J., Morris G. | Accounting Forum | 2013 |
| 137 | Do Auditor-Provided Tax Services Enhance Or Impair The Value Relevance Of Earnings? | Cook K. | Journal of the American Taxation Association | 2013 |
| 138 | Schedule Utp: Stock Price Reaction And Economic Consequences | Abernathy J.L., Davenport S.A., Rapley E.T. | Journal of the American Taxation Association | 2013 |
| 139 | Religiosity And Tax Avoidance | Boone J., Khurana I., Raman K. | Journal of the American Taxation Association | 2013 |
| 140 | Do Publicly Disclosed Tax Reserves Tell Us About Privately Disclosed Tax Shelter Activity? | Lisowsky P., Robinson L., Schmidt A. | Journal of Accounting Research | 2013 |
| 141 | Exploring The Role Delaware Plays As A Domestic Tax Haven | Dyreng S.D., Lindsey B.P., Thornock J.R. | Journal of Financial Economics | 2013 |

(continued)

**Table A.1** (continued)

| N | Title | Authors | Journal | Year |
|---|---|---|---|---|
| 142 | Labor Unions And Tax Aggressiveness | Ching Leung W.S., Chyz J.A., Meng Rui O., Zhen Li O. | Journal of Financial Economics | 2013 |
| 143 | Multinational Transfer Pricing: A Transaction Cost And Resource Based View | Cecchini M., Leitch R., Strobel C. | Journal of Accounting Literature | 2013 |
| 144 | The Missing Wealth Of Nations: Are Europe And The U.S. Net Debtors Or Net Creditors? | Zucman G. | Quarterly Journal of Economics | 2013 |
| 145 | Earnings, Institutional Investors, Tax Avoidance, And Firm Value: Evidence From Taiwan | Chang L.L., Hsiao F.D., Tsai Y.C. | Journal of International Accounting, Auditing and Taxation | 2013 |
| 146 | Tax Aggressiveness And Accounting Fraud | Lennox C., Lisowsky P., Pittman J. | Journal of Accounting Research | 2013 |
| 147 | Challenging The Reliability Of Comparables Under Profit-Based Transfer Pricing Methods | Emmanuel C., Mura A., Vallascas F. | Accounting and Business Research | 2013 |
| 148 | The Relationship Between The Management Of Book Income And Taxable Income Under A Moderate Level Of Book-Tax Conformity | Chen E., Gavious I., Yosef R. | Journal of Accounting, Auditing and Finance | 2013 |
| 149 | Tax Bunching, Income Shifting And Self-Employment | le Maire D., Schjerning B. | Journal of Public Economics | 2013 |
| 150 | The Insurance Hypothesis: An Examination Of Kpmg'S Audit Clients Around The Investigation And Settlement Of The Tax Shelter Case | Brown D., Shu S., Soo B., Trompeter G. | Auditing | 2013 |
| 151 | Excise Tax Avoidance: The Case Of State Cigarette Taxes | DeCicca P., Kenkel D., Liu F. | Journal of Health Economics | 2013 |
| 152 | Firm And Investor Responses To Uncertain Tax Benefit Disclosure Requirements | Robinson L., Schmidt A. | Journal of the American Taxation Association | 2013 |
| 153 | Corporate Tax Avoidance And The Timeliness Of Annual Earnings Announcements | Crabtree A., Kubick T. | Review of Quantitative Finance and Accounting | 2014 |
| 154 | The End Of Bank Secrecy? An Evaluation Of The G20 Tax Haven Crackdown | Johannesen N., Zucman G. | American Economic Journal: Economic Policy | 2014 |
| 155 | Does Corporate Tax Aggressiveness Influence Audit Pricing? | Donohoe M., Robert Knechel W. | Contemporary Accounting Research | 2014 |
| 156 | Coordinating Production And Marketing With Dynamic Transfer Prices | Dockner E.J., Fruchter G.E. | Production and Operations Management | 2014 |

(continued)

**Table A.1** (continued)

| N | Title | Authors | Journal | Year |
|---|---|---|---|---|
| 157 | Consumer Reactions To Corporate Tax Strategies: Effects On Corporate Reputation And Purchasing Behavior | Hardeck I., Hertl R. | Journal of Business Ethics | 2014 |
| 158 | A Journey From A Corruption Port To A Tax Haven | Hebous S., Lipatov V. | Journal of Comparative Economics | 2014 |
| 159 | A Model Of Multinational Income Shifting And An Application To Tax Planning With E-Commerce | Carnaghan C., Klassen K.J., Laplante S.K. | Journal of the American Taxation Association | 2014 |
| 160 | The Effect Of Mandatory Financial Statement Disclosures Of Tax Uncertainty On Tax Reporting And Collections: The Case Of Fin 48 And Multistate Tax Avoidance | Gupta S., Mills L., Towery E. | Journal of the American Taxation Association | 2014 |
| 161 | Optimal Software Reuse In Incremental Software Development: A Transfer Pricing Approach | Ceran Y., Dawande M., Liu D., Mookerjee V. | Management Science | 2014 |
| 162 | Beauty Is In The Eye Of The Beholder: The Effect Of Corporate Tax Avoidance On The Cost Of Bank Loans | Hasan I., Hoi C., Wu Q., Zhang H. | Journal of Financial Economics | 2014 |
| 163 | Are Female Cfos Less Tax Aggressive? Evidence From Tax Aggressiveness | Francis B.B., Hasan I., Wu Q., Yan M. | Journal of the American Taxation Association | 2014 |
| 164 | Taxing Audit Markets And Reputation: An Examination Of The U.S. Tax Shelter Controversy | Higgs J.L., Huang I., Incardona J., Kannan Y., Premuroso R. | Journal of International Accounting, Auditing and Taxation | 2014 |
| 165 | Honor Among Tax Havens | Bucovetsky S. | Journal of Public Economics | 2014 |
| 166 | Bilateral Internal Debt Financing And Tax Planning Of Multinational Firms | Overesch M., Wamser G. | Review of Quantitative Finance and Accounting | 2014 |
| 167 | Optimal Taxation Of Top Labor Incomes: A Tale Of Three Elasticities | Piketty T., Saez E., Stantcheva S. | American Economic Journal: Economic Policy | 2014 |
| 168 | Tax Evasion And Swiss Bank Deposits | Johannesen N. | Journal of Public Economics | 2014 |
| 169 | Tax Incentives And The Demand For Private Health Insurance | Stavrunova O., Yerokhin O. | Journal of Health Economics | 2014 |
| 170 | Tax Avoidance With Cross-Border Hybrid Instruments | Johannesen N. | Journal of Public Economics | 2014 |
| 171 | Relative Valuation Of Alternative Methods Of Tax Avoidance | Brown J. | Journal of the American Taxation Association | 2014 |

(continued)

**Table A.1** (continued)

| N | Title | Authors | Journal | Year |
|---|---|---|---|---|
| 172 | Relative Valuation Of Alternative Methods Of Tax Avoidance | Inger K. | Journal of the American Taxation Association | 2014 |
| 173 | Public Versus Private Firm Responses To The Tax Rate Reduction In China | Lin K.Z., Mills L.F., Zhang F. | Journal of the American Taxation Association | 2014 |
| 174 | Investment Opportunity Sets, Operating Uncertainty, And Capital Market Pressure: Determinants Of Investments In Tax Shelter Activities? | McGuire S., Omer T., Wilde J. | Journal of the American Taxation Association | 2014 |
| 175 | Transfer Pricing As A Tax Compliance Risk | Jost S., Pfaffermayr M., Winner H. | Accounting and Business Research | 2014 |
| 176 | Ethics In Tax Practice: A Study Of The Effect Of Practitioner Firm Size | Doyle E., Frecknall-Hughes J., Summers B. | Journal of Business Ethics | 2014 |
| 177 | Dual Class Ownership And Tax Avoidance | McGuire S.T., Wang D., Wilson R.J. | Accounting Review | 2014 |
| 178 | The Curious Case Of Corporate Tax Avoidance: Is It Socially Irresponsible? | Dowling G.R. | Journal of Business Ethics | 2014 |
| 179 | Ceo After-Tax Compensation Incentives And Corporate Tax Avoidance | Gaertner F.B. | Contemporary Accounting Research | 2014 |
| 180 | Transfer Pricing And Sourcing Strategies For Multinational Firms | Debo L., Gavirneni S., Shunko M. | Production and Operations Management | 2014 |
| 181 | The Reputational Costs Of Tax Avoidance | Gallemore J., Maydew E.L., Thornock J.R. | Contemporary Accounting Research | 2014 |
| 182 | International Transfer Pricing | Sansing R. | Foundations and Trends in Accounting | 2015 |
| 183 | An Empirical Analysis Of The Changes In Tax Audit Focus On International Transfer Pricing | Chan K., Lo A., Mo P. | Journal of International Accounting, Auditing and Taxation | 2015 |
| 184 | Taxpayer Search For Information: Implications For Rational Attention | Hoopes J., Reck D., Slemrod J. | American Economic Journal: Economic Policy | 2015 |
| 185 | Taking The Long Way Home: U.S. Tax Evasion And Offshore Investments In U.S. Equity And Debt Markets | Hanlon M., Maydew E.L., Thornock J.R. | Journal of Finance | 2015 |

(continued)

**Table A.1** (continued)

| N | Title | Authors | Journal | Year |
|---|---|---|---|---|
| 186 | Financial Derivatives In Corporate Tax Avoidance: A Conceptual Perspective | Donohoe M.P. | Journal of the American Taxation Association | 2015 |
| 187 | Is Corporate Social Responsibility Performance Associated With Tax Avoidance? | Lanis R., Richardson G. | Journal of Business Ethics | 2015 |
| 188 | Financial Accounting Effects Of Tax Aggressiveness: Contracting And Measurement | de Waegenaere A., Sansing R., Wielhouwer J.L. | Contemporary Accounting Research | 2015 |
| 189 | The Association Between Book-Tax Conformity And Earnings Management | Blaylock B., Gaertner F., Shevlin T. | Review of Accounting Studies | 2015 |
| 190 | The Taxation Of Foreign Profits: A Unified View | Devereux M.P., Fuest C., Lockwood B. | Journal of Public Economics | 2015 |
| 191 | The Political Economics Of Redistribution, Inequality And Tax Avoidance | Bethencourt C., Kunze L. | Public Choice | 2015 |
| 192 | The Influence Of A Firm'S Business Strategy On Its Tax Aggressiveness | Higgins D., Omer T.C., Phillips J.D. | Contemporary Accounting Research | 2015 |
| 193 | Manufacturing Capacity Decisions With Demand Uncertainty And Tax Cross-Crediting | Hsu V.N., Hu Q., Xiao W. | Manufacturing and Service Operations Management | 2015 |
| 194 | The Hand Of Accounting And Accountancy Firms In Deepening Income And Wealth Inequalities And The Economic Crisis: Some Evidence | Sikka P. | Critical Perspectives on Accounting | 2015 |
| 195 | Does Book-Tax Conformity Deter Opportunistic Book And Tax Reporting? An International Analysis | Tang T.Y.H. | European Accounting Review | 2015 |
| 196 | Audit Partner Public-Client Specialisation And Client Abnormal Accruals | Ittonen K., Johnstone K., Myllymäki E.R. | European Accounting Review | 2015 |
| 197 | Do Industry-Specialist Auditors Influence Stock Price Crash Risk? | Robin A.J., Zhang H. | Auditing | 2015 |
| 198 | Corporate Social Responsibility, Tax Avoidance, And Earnings Performance | Watson L. | Journal of the American Taxation Association | 2015 |
| 199 | Has Regulation Of Charitable Foundations Thrown The Baby Out With The Bath Water? | Marx B.M. | Journal of Public Economics | 2015 |
| 200 | Book-Tax Differences And Future Earnings Changes | Jackson M. | Journal of the American Taxation Association | 2015 |

(continued)

**Table A.1** (continued)

| N | Title | Authors | Journal | Year |
|---|---|---|---|---|
| 201 | Taxes And Financial Constraints: Evidence From Linguistic Cues | Law K.K.F., Mills L.F. | Journal of Accounting Research | 2015 |
| 202 | The Impact Of Controlled Foreign Company Legislation On Real Investments Abroad. A Multi-Dimensional Regression Discontinuity Design | Egger P., Wamser G. | Journal of Public Economics | 2015 |
| 203 | Smoothness And The Value Relevance Of Taxable Income | Mayberry M., McGuire S., Omer T. | Journal of the American Taxation Association | 2015 |
| 204 | The Influence Of Governance Infrastructure And Corporate Governance On Profit Shifting | George R., Sugathan A. | Journal of International Business Studies | 2015 |
| 205 | Labor Tax Avoidance And Its Determinants: The Case Of Mafia Firms In Italy | Argilés-Bosch J.M., Ravenda D., Valencia-Silva M.M. | Journal of Business Ethics | 2015 |
| 206 | The Persistence Of Book-Tax Differences | Abdul Wahab N.S., Holland K. | British Accounting Review | 2015 |
| 207 | For Logistical Reasons Only? A Case Study Of Tax Planning And Corporate Social Responsibility Reporting | Laine M., Ylönen M. | Critical Perspectives on Accounting | 2015 |
| 208 | Top Management Conservatism And Corporate Risk Strategies: Evidence From Managers' Personal Political Orientation And Corporate Tax Avoidance | Boivie S., Christensen D.M., Dhaliwal D.S., Graffin S.D. | Strategic Management Journal | 2015 |
| 209 | Is Tax Avoidance Associated With Economically Significant Rent Extraction Among U.S. Firms? | Blaylock B.S. | Contemporary Accounting Research | 2016 |
| 210 | Firm-Level Tournament Incentives And Corporate Tax Aggressiveness | Kubick T.R., Masli A.N.S. | Journal of Accounting and Public Policy | 2016 |
| 211 | Tax-Induced Distortions Of Effort And Compensation In A Principal-Agent Setting | Martini J., Niemann R., Simons D. | Journal of International Accounting, Auditing and Taxation | 2016 |
| 212 | Auditor-Provided Tax Services And Stock Price Crash Risk | Habib A., Hasan M. | Accounting and Business Research | 2016 |
| 213 | A Comparison Of The Tax-Motivated Income Shifting Of Multinationals In Territorial And Worldwide Countries | Markle K. | Contemporary Accounting Research | 2016 |
| 214 | Discussion Of "Corporate Political Connections And Tax Aggressiveness" | Faccio M. | Contemporary Accounting Research | 2016 |

(continued)

**Table A.1** (continued)

| N | Title | Authors | Journal | Year |
|---|---|---|---|---|
| 215 | Discussion Of Ceo Narcissism And Corporate Tax Sheltering | Brown J. | Journal of the American Taxation Association | 2016 |
| 216 | Corporate Political Connections And Tax Aggressiveness | Kim C., Zhang L. | Contemporary Accounting Research | 2016 |
| 217 | Ceo Narcissism And Corporate Tax Sheltering | Olsen K.J., Stekelberg J. | Journal of the American Taxation Association | 2016 |
| 218 | The Economic Consequences Of Tax Service Provider Sanctions: Evidence From Kpmg'S Deferred Prosecution Agreement | Finley A.R., Stekelberg J. | Journal of the American Taxation Association | 2016 |
| 219 | Compensation In The Post-Fin 48 Period: The Case Of Contracting On Tax Performance And Uncertainty | Brown J.L., Drake K.D., Martin M.A. | Contemporary Accounting Research | 2016 |
| 220 | Public Pressure And Corporate Tax Behavior | Dyreng S.D., Hoopes J.L., Wilde J.H. | Journal of Accounting Research | 2016 |
| 221 | General Counsel Prominence And Corporate Tax Policy | Abernathy J.L., Kubick T.R., Masli A. | Journal of the American Taxation Association | 2016 |
| 222 | Risk Transfer Valuation In Advance Pricing Agreements Between Multinational Enterprises And Tax Authorities | Afik Z., Lahav Y. | Journal of Accounting, Auditing and Finance | 2016 |
| 223 | Tax Avoidance And The Implications Of Weak Internal Controls | Bauer A.M. | Contemporary Accounting Research | 2016 |
| 224 | Allowing Firms To Choose Between Separate Accounting And Formula Apportionment Taxation | Gresik T. | Journal of Public Economics | 2016 |
| 225 | Taxing Multinationals In The Presence Of Internal Capital Markets | Koethenbuerger M., Stimmelmayr M. | Journal of Public Economics | 2016 |
| 226 | Dark Matter, Black Holes And Old-Fashioned Exploitation: Transnational Corporations And The Us Economy | Ali M. | Cambridge Journal of Economics | 2016 |
| 227 | Convergence In Corporate Statutory Tax Rates In The Asian And Pacific Economies | Chen Y., Cuestas J.C., Regis P.J. | International Journal of Finance and Economics | 2016 |
| 228 | Transfer Pricing By Multinational Firms: New Evidence From Foreign Firm Ownerships | Cristea A., Nguyen D. | American Economic Journal: Economic Policy | 2016 |
| 229 | Silence Is Golden? Evidence From Disclosing Related-Party Transactions In China | Lo A., Wong R. | Journal of Accounting and Public Policy | 2016 |

(continued)

**Table A.1** (continued)

| N | Title | Authors | Journal | Year |
|---|---|---|---|---|
| 230 | Do Investors Use Prior Tax Avoidance When Pricing Tax Loss Carryforwards? | McGuire S.T., Neuman S.S., Olson A.J., Omer T.C. | Journal of the American Taxation Association | 2016 |
| 231 | Dividend Taxes And Income Shifting | Alstadsæter A., Jacob M. | Scandinavian Journal of Economics | 2016 |
| 232 | Multinational Firms And Tax Havens | Gumpert A., Hines J.R., Schnitzer M. | Review of Economics and Statistics | 2016 |
| 233 | Public Provision Of Private Goods, Self-Selection, And Income Tax Avoidance | Blomquist S., Christiansen V., Micheletto L. | Scandinavian Journal of Economics | 2016 |
| 234 | Coordination And The Fight Against Tax Havens | Konrad K., Stolper T. | Journal of International Economics | 2016 |
| 235 | Corporate Patents, R&D Success, And Tax Avoidance | Gao L., Yang L., Zhang J. | Review of Quantitative Finance and Accounting | 2016 |
| 236 | Relation Between Auditor Quality And Tax Aggressiveness: Implications Of Cross-Country Institutional Differences | Kanagaretnam K., Lee J., Lim C., Lobo G. | Auditing | 2016 |
| 237 | Is Transfer Pricing Strictness Deterring Profit Shifting Within Multinationals? Empirical Evidence From Europe | Marques M., Pinho C. | Accounting and Business Research | 2016 |
| 238 | Do Corporate Tax Cuts Increase Investments? | Dobbins L., Jacob M. | Accounting and Business Research | 2016 |
| 239 | Discussion Of "Target'S Tax Shelter Participation And Takeover Premiums" | Shevlin T. | Contemporary Accounting Research | 2016 |
| 240 | On The Implications Of Introducing Cross-Border Loss-Offset In The European Union | Kalamov Z.Y., Runkel M. | Journal of Public Economics | 2016 |
| 241 | Targets' Tax Shelter Participation And Takeover Premiums | Chow T., Klassen K., Liu Y. | Contemporary Accounting Research | 2016 |
| 242 | Taxing The Rich | Landier A., Plantin G. | Review of Economic Studies | 2017 |
| 243 | Anticipation, Tax Avoidance, And The Price Elasticity Of Gasoline Demand | Coglianese J., Davis L., Kilian L., Stock J. | Journal of Applied Econometrics | 2017 |
| 244 | Debt Maturity And Tax Avoidance | Platikanova P. | European Accounting Review | 2017 |
| 245 | Customer–Supplier Relationships And Corporate Tax Avoidance | Cen L., Maydew E.L., Zhang L., Zuo L. | Journal of Financial Economics | 2017 |

(continued)

**Table A.1** (continued)

| N | Title | Authors | Journal | Year |
|---|---|---|---|---|
| 246 | The Role Of Informal Controls And A Bargaining Opponent'S Emotions On Transfer Pricing Judgments | Bhattacharjee S., Moreno K.K. | Contemporary Accounting Research | 2017 |
| 247 | The Three Hurdles Of Tax Planning: How Business Context, Aims Of Tax Planning, And Tax Manager Power Affect Tax Expense | Feller A., Schanz D. | Contemporary Accounting Research | 2017 |
| 248 | An Examination Of Reputational Costs And Tax Avoidance: Evidence From Firms With Valuable Consumer Brands | Austin C.R., Wilson R.J. | Journal of the American Taxation Association | 2017 |
| 249 | Transfer Pricing: Strategies, Practices, And Tax Minimization | Klassen K.J., Lisowsky P., Mescall D. | Contemporary Accounting Research | 2017 |
| 250 | Income Statement Reporting Discretion Allowed By Fin 48: Interest And Penalty Expense Classification | Abernathy J.L., Beyer B., Gross A.D., Rapley E.T. | Journal of the American Taxation Association | 2017 |
| 251 | The Executive Compensation Implications Of The Tax Component Of Earnings | Hansen V., Lopez T.J., Reitenga A. | Review of Quantitative Finance and Accounting | 2017 |
| 252 | Profit Shifting Of U.S. Multinationals | Dowd T., Landefeld P., Moore A. | Journal of Public Economics | 2017 |
| 253 | Overconfidence, Ceo Awards, And Corporate Tax Aggressiveness | Kubick T., Lockhart G. | Journal of Business Finance and Accounting | 2017 |
| 254 | Controlling Shareholders' Incentive And Corporate Tax Avoidance: A Natural Experiment In China | Li O., Liu H., Ni C. | Journal of Business Finance and Accounting | 2017 |
| 255 | Beyond Tax Avoidance: Offshore Firms' Institutional Environment And Financial Reporting Quality* | Durnev A., Li T., Magnan M. | Journal of Business Finance and Accounting | 2017 |
| 256 | Deciphering Tax Avoidance: Evidence From Credit Rating Disagreements | Bonsall S., Koharki K., Watson L. | Contemporary Accounting Research | 2017 |
| 257 | Tax Havens, Tax Evasion And Tax Information Exchange Agreements In The Oecd | Kemme D., Parikh B., Steigner T. | European Financial Management | 2017 |
| 258 | Social Capital And Debt Contracting: Evidence From Bank Loans And Public Bonds | Hasan I., Hoi C.K., Wu Q., Zhang H. | Journal of Financial and Quantitative Analysis | 2017 |
| 259 | Does Social Capital Matter In Corporate Decisions? Evidence From Corporate Tax Avoidance | Hasan I., Hoi C.K.S., Wu Q., Zhang H. | Journal of Accounting Research | 2017 |

(continued)

**Table A.1** (continued)

| N | Title | Authors | Journal | Year |
|---|---|---|---|---|
| 260 | Changes In Corporate Effective Tax Rates Over The Past 25 Years | Dyreng S.D., Hanlon M., Maydew E.L., Thornock J.R. | Journal of Financial Economics | 2017 |
| 261 | Agency Costs Of Moving To Tax Havens: Evidence From Cross-Border Merger Premia | Col B. | Corporate Governance: An International Review | 2017 |
| 262 | Tax Avoidance And Ex Ante Cost Of Capital | Cook K., Moser W., Omer T. | Journal of Business Finance and Accounting | 2017 |
| 263 | Anonymous Shell Companies: A Global Audit Study And Field Experiment In 176 Countries | Allred B.B., Findley M.G., Nielson D., Sharman J.C. | Journal of International Business Studies | 2017 |
| 264 | To Shift Or Not To Shift? Intertemporal Income Shifting As A Response To The Risk Capital Allowance Introduction In Belgium | Andries K., Cools M., Van Uytbergen S. | European Accounting Review | 2017 |
| 265 | Does A Firm'S Life Cycle Explain Its Propensity To Engage In Corporate Tax Avoidance? | Al-Hadi A., Hasan M.M., Richardson G., Taylor G. | European Accounting Review | 2017 |
| 266 | Bank-Owned Life Insurance And Bank Risk | Davidson T. | Financial Review | 2017 |
| 267 | Immobilizing Corporate Income Shifting: Should It Be Safe To Strip In The Harbor? | Gresik T.A., Schindler D., Schjelderup G. | Journal of Public Economics | 2017 |
| 268 | Does Schedule Utp Have Uniform Long-Run Effects On Corporate Tax Planning? | Honaker K., Sharma D. | Journal of the American Taxation Association | 2017 |
| 269 | Equity Compensation And Tax Avoidance: Disentangling Managerial Incentives From Tax Benefits And Reexamining The Effect Of Shareholder Rights | Seidman J.K., Stomberg B. | Journal of the American Taxation Association | 2017 |
| 270 | Board Of Director Gender And Corporate Tax Aggressiveness: An Empirical Analysis | Lanis R., Richardson G., Taylor G. | Journal of Business Ethics | 2017 |
| 271 | Ceo Inside Debt Incentives And Corporate Tax Sheltering | Chi S., Huang S., Sanchez J. | Journal of Accounting Research | 2017 |
| 272 | Are Socially Responsible Firms Less Likely To Expatriate? An Examination Of Corporate Inversions | Huang H.H., Sun L., Yu T.R. | Journal of the American Taxation Association | 2017 |
| 273 | The Market Valuation Of The Permanent Book-To-Tax Differences Generated By Stock-Based Compensation Awards | Brushwood J.D., Johnston D., Kutcher L. | Journal of the American Taxation Association | 2017 |

(continued)

Table A.1 (continued)

| N | Title | Authors | Journal | Year |
|---|---|---|---|---|
| 274 | Tax-Driven Wealth Chains: A Multiple Case Study Of Tax Avoidance In The Finnish Mining Sector | Finér L., Ylönen M. | Critical Perspectives on Accounting | 2017 |
| 275 | Does Information Asymmetry Affect Corporate Tax Aggressiveness? | Chen T., Lin C. | Journal of Financial and Quantitative Analysis | 2017 |
| 276 | The Role Of Managerial Ability In Corporate Tax Avoidance | Koester A., Shevlin T., Wangerin D. | Management Science | 2017 |
| 277 | Multinationals' Profit Response To Tax Differentials: Effect Size And Shifting Channels | Heckemeyer J., Overesch M. | Canadian Journal of Economics | 2017 |
| 278 | An Examination Of The Political Salience Of Corporate Tax Avoidance: A Case Study Of The Tax Justice Network | Dallyn S. | Accounting Forum | 2017 |
| 279 | Accounting And Taxation: Conjoined Twins Or Separate Siblings? | Sikka P. | Accounting Forum | 2017 |
| 280 | Evaluating The Effects Of Ace Systems On Multinational Debt Financing And Investment | Hebous S., Ruf M. | Journal of Public Economics | 2017 |
| 281 | An Examination Of Ethical Influences On The Work Of Tax Practitioners | Doyle E., Frecknall-Hughes J., Moizer P., Summers B. | Journal of Business Ethics | 2017 |
| 282 | Taxes, Informality And Income Shifting: Evidence From A Recent Pakistani Tax Reform | Waseem M. | Journal of Public Economics | 2018 |
| 283 | Outside Directors, Corporate Social Responsibility Performance, And Corporate Tax Aggressiveness: An Empirical Analysis | Lanis R., Richardson G. | Journal of Accounting, Auditing and Finance | 2018 |
| 284 | Tax Havens And Disclosure Aggregation | Akamah H., Hope O.K., Thomas W.B. | Journal of International Business Studies | 2018 |
| 285 | Aggressive Tax Avoidance: A Conundrum For Stakeholders, Governments, And Morality | Payne D.M., Raiborn C.A. | Journal of Business Ethics | 2018 |
| 286 | Conflicting Transfer Pricing Incentives And The Role Of Coordination | Blouin J., Robinson L., Seidman J. | Contemporary Accounting Research | 2018 |
| 287 | A Study Of Perceived Ethicality Of Low Corporate Effective Tax Rates | DeZoort F.T., Pollard T.J., Schnee E.J. | Accounting Horizons | 2018 |

(continued)

**Table A.1** (continued)

| N | Title | Authors | Journal | Year |
|---|---|---|---|---|
| 288 | Are Corporate Risk-Taking Practices Indicative Of Aggressive Reporting Practices? | Frank M.M., Lynch L.J., Rego S.O., Zhao R. | Journal of the American Taxation Association | 2018 |
| 289 | An Examination Of The Impact Of Tax Avoidance On The Readability Of Tax Footnotes | Fan W.P., Inger K.K., Meckfessel M.D., Zhou M.J. | Journal of the American Taxation Association | 2018 |
| 290 | An Analysis Of The Corporate Income Tax Policy Of Less Developed Countries | Baker P.L. | Scandinavian Journal of Economics | 2018 |
| 291 | Quantifying Loss-Averse Tax Manipulation | Rees-Jones A. | Review of Economic Studies | 2018 |
| 292 | Corporate Tax Havens And Transparency | Bennedsen M., Zeume S. | Review of Financial Studies | 2018 |
| 293 | Overconfidence And Tax Avoidance: The Role Of Ceo And Cfo Interaction | Demirkan S., Hsieh T.S., Wang Z. | Journal of Accounting and Public Policy | 2018 |
| 294 | Parent-Subsidiary Investment Layers And Audit Fees | Gul F.A., Hsu A.W.H., Liu S.H.T. | Journal of Accounting, Auditing and Finance | 2018 |
| 295 | How Does Transfer Pricing Risk Affect Premiums In Cross-Border Mergers And Acquisitions? | Klassen K.J., Mescall D. | Contemporary Accounting Research | 2018 |
| 296 | Who Owns The Wealth In Tax Havens? Macro Evidence And Implications For Global Inequality | Alstadsæter A., Johannesen N., Zucman G. | Journal of Public Economics | 2018 |
| 297 | Do Analysts' Cash Flow Forecasts Encourage Managers To Improve The Firm'S Cash Flows? Evidence From Tax Planning | Ayers B.C., Call A.C., Schwab C.M. | Contemporary Accounting Research | 2018 |
| 298 | Do Analysts Matter For Corporate Tax Planning? Evidence From A Natural Experiment | Chen N.X., Chiu P.C., Shevlin T. | Contemporary Accounting Research | 2018 |
| 299 | Cross-Country Evidence On The Role Of Independent Media In Constraining Corporate Tax Aggressiveness | Kanagaretnam K., Lee J., Lim C.Y., Lobo G.J. | Journal of Business Ethics | 2018 |
| 300 | Offshore Expertise For Onshore Companies: Director Connections To Island Tax Havens And Corporate Tax Policy | Jiang C., Kubick T.R., Miletkov M.K., Wintoki M.B. | Management Science | 2018 |
| 301 | Formula Apportionment: Factor Allocation And Tax Avoidance | Eichfelder S., Hechtner F., Hundsdoerfer J. | European Accounting Review | 2018 |
| 302 | Tax Avoidance As A Sustainability Problem | Bird R., Davis-Nozemack K. | Journal of Business Ethics | 2018 |

(continued)

**Table A.1** (continued)

| N | Title | Authors | Journal | Year |
|---|---|---|---|---|
| 303 | Perspectives On Corporate Tax Planning: Observations From The Past Decade | Wilde J., Wilson R. | Journal of the American Taxation Association | 2018 |
| 304 | Corporate Governance, Tax Avoidance, And Financial Constraints | Bayar O., Huseynov F., Sardarli S. | Financial Management | 2018 |
| 305 | Double Tax Discrimination To Attract Fdi And Fight Profit Shifting: The Role Of Cfc Rules | Haufler A., Mardan M., Schindler D. | Journal of International Economics | 2018 |
| 306 | Optimal Pricing And Inventory Planning With Charitable Donations | Chu L.Y., Li G., Rusmevichientong P. | Manufacturing and Service Operations Management | 2018 |
| 307 | Internal Information Quality And Tax-Motivated Income Shifting | McGuire S., Rane S., Weaver C. | Journal of the American Taxation Association | 2018 |
| 308 | Corporate Tax Compliance: Is A Change Towards Trust-Based Tax Strategies Justified? | Goslinga S., Siglé M., Speklé R., van der Hel L., Veldhuizen R. | Journal of International Accounting, Auditing and Taxation | 2018 |
| 309 | Tax Avoidance, Financial Experts On The Audit Committee, And Business Strategy | Hsu P.H., Moore J.A., Neubaum D.O. | Journal of Business Finance and Accounting | 2018 |
| 310 | Executive Cash Compensation And Tax Aggressiveness Of Chinese Firms | Huang W., Shen Y., Ying T. | Review of Quantitative Finance and Accounting | 2018 |
| 311 | Multinational Tax Avoidance: Virtue Ethics And The Role Of Accountants | West A. | Journal of Business Ethics | 2018 |
| 312 | The Information Content Of Tax Expense: A Discount Rate Explanation | Henry E. | Contemporary Accounting Research | 2018 |
| 313 | Business Groups And Tax Havens | Su W., Tan D. | Journal of Business Ethics | 2018 |
| 314 | Tax Avoidance Over Time: A Comparison Of European And U.S. Firms | Thomsen M., Watrin C. | Journal of International Accounting, Auditing and Taxation | 2018 |
| 315 | An Evaluation Of Alternative Market-Based Transfer Prices | Johnson N.B., Loeffler C., Pfeiffer T. | Contemporary Accounting Research | 2018 |
| 316 | Tax Avoidance, Managerial Ability, And Investment Efficiency | Khurana I., Moser W., Raman K. | Abacus | 2018 |
| 317 | Do Political Connections Weaken Tax Enforcement Effectiveness? | Li Y., Lin K.Z., Mills L.F., Zhang F. | Contemporary Accounting Research | 2018 |
| 318 | Us Tax Reform: A Multilateral Success? | Herzfeld M. | British Tax Review | 2019 |

(continued)

**Table A.1** (continued)

| N | Title | Authors | Journal | Year |
|---|---|---|---|---|
| 319 | Challenges To Multilateralism In International Taxation: A Tale Of Two Measures | Tandon S. | British Tax Review | 2019 |
| 320 | Multilateral Dynamics In Bilateral Settings: Back To Realpolitik | Antón R.G. | British Tax Review | 2019 |
| 321 | The Effect Of Estate Tax Change On The Controlling Shareholding Structure And Corporate Value Of Family Firms | Liao C., Yeh Y. | Corporate Governance: An International Review | 2019 |
| 322 | 'Paying Taxes Is Losing Money': A Qualitative Study On Institutional Logics In The Tax Consultancy Field In Romania | Apostol O., Pop A. | Critical Perspectives on Accounting | 2019 |
| 323 | Will Destination-Based Taxes Be Fully Exploited When Available? An Application To The U.S. Commodity Tax System | Agrawal D.R., Mardan M. | Journal of Public Economics | 2019 |
| 324 | Trade Liberalization And Corporate Income Tax Avoidance | Gao W., Lu Y., Shi X. | Economic Inquiry | 2019 |
| 325 | Does Tax Risk Affect Investor Valuation Of Tax Avoidance? | Drake K.D., Lusch S.J., Stekelberg J. | Journal of Accounting, Auditing and Finance | 2019 |
| 326 | The Effect Of Investing Abroad On Investment At Home: On The Role Of Technology, Tax Savings, And Internal Capital Markets | Goldbach S., Nagengast A.J., Steinmüller E., Wamser G. | Journal of International Economics | 2019 |
| 327 | An Estimable Model Of Income Redistribution In A Federation: Musgrave Meets Oates | Milligan K., Smart M. | American Economic Journal: Economic Policy | 2019 |
| 328 | Did The 2005 Deferred Prosecution Agreement Adversely Impact Kpmg'S Audit Practice? | Baugh M., Boone J.P., Khurana I.K., Raman K.K. | Auditing | 2019 |
| 329 | Going To Haven? Corporate Social Responsibility And Tax Avoidance | Col B., Patel S. | Journal of Business Ethics | 2019 |
| 330 | The Potential Of Tax Surprises To Affect Measures Of Tax Avoidance And Researchers' Inferences | Austin C.R. | Journal of the American Taxation Association | 2019 |
| 331 | The Complementarity Between Tax Avoidance And Manager Diversion: Evidence From Tax Haven Firms | Atwood T.J., Lewellen C. | Contemporary Accounting Research | 2019 |
| 332 | Income Shifting Using A Cost-Sharing Arrangement | De Simone L., Sansing R. | Journal of the American Taxation Association | 2019 |
| 333 | Banks As Tax Planning Intermediaries | Gallemore J., Gipper B., Maydew E. | Journal of Accounting Research | 2019 |

(continued)

**Table A.1** (continued)

| N | Title | Authors | Journal | Year |
|---|---|---|---|---|
| 334 | The Interplay Between Mandatory Country-By-Country Reporting, Geographic Segment Reporting, And Tax Havens: Evidence From The European Union | Brown R.J., Jorgensen B.N., Pope P.F. | Journal of Accounting and Public Policy | 2019 |
| 335 | The Determinants And Consequences Of Tax Audits: Some Evidence From China | Li W., Pittman J.A., Wang Z.T. | Journal of the American Taxation Association | 2019 |
| 336 | Corporate Tax Aggressiveness And Insider Trading | Chung S., Goh B., Lee J., Shevlin T. | Contemporary Accounting Research | 2019 |
| 337 | Limits Of Tax Regulation: Evidence From Strategic R&D Classification And The R&D Tax Credit | Laplante S.K., Skaife H.A., Swenson L.A., Wangerin D.D. | Journal of Accounting and Public Policy | 2019 |
| 338 | Joint Audit Engagements And Client Tax Avoidance: Evidence From The Italian Statutory Audit Regime | Bianchi P.A., Falsetta D., Minutti-Meza M., Weisbrod E. | Journal of the American Taxation Association | 2019 |
| 339 | The Impact Of Large Tax Settlement Favorability On Firms' Subsequent Tax Avoidance | Finley A. | Review of Accounting Studies | 2019 |
| 340 | Do Auditors Constrain Intertemporal Income Shifting In Private Companies? | Höglund H., Sundvik D. | Accounting and Business Research | 2019 |
| 341 | Corporate Profitability And Effective Tax Rate: The Enforcement Effect Of Large Taxpayer Units | Tennant S.N., Tracey M.R. | Accounting and Business Research | 2019 |
| 342 | Tone-At-The-Top Lessons From Abrahamic Justice | Fischer D., Friedman H.H. | Journal of Business Ethics | 2019 |
| 343 | Taxing Multinationals Beyond Borders: Financial And Locational Responses To Cfc Rules | Clifford S. | Journal of Public Economics | 2019 |
| 344 | Profit Shifting And Investment Effects: The Implications Of Zero-Taxable Profits | Koethenbuerger M., Mardan M., Stimmelmayr M. | Journal of Public Economics | 2019 |
| 345 | The Corrosive Effect Of Offshore Financial Centers On Multinational Firms' Disclosure Strategy | Ben Amar W., He L., Li T., Magnan M. | European Accounting Review | 2019 |
| 346 | Transparency, Information Shocks, And Tax Avoidance | Kerr J.N. | Contemporary Accounting Research | 2019 |
| 347 | Taxes And Firm Size: Political Cost Or Political Power? | Belz T., Steffens C., von Hagen D. | Journal of Accounting Literature | 2019 |
| 348 | The Impact Of State Taxes On Pass-Through Businesses: Evidence From The 2012 Kansas Income Tax Reform | DeBacker J., Heim B.T., Ramnath S.P., Ross J.M. | Journal of Public Economics | 2019 |

(continued)

**Table A.1** (continued)

| N | Title | Authors | Journal | Year |
|---|---|---|---|---|
| 349 | Sharing Corporate Tax Knowledge With External Advisers | Hasseldine J., Holland K., van der Rijt P. | Accounting and Business Research | 2019 |
| 350 | Using Irs Data To Identify Income Shifting To Foreign Affiliates | De Simone L., Mills L.F., Stomberg B. | Review of Accounting Studies | 2019 |
| 351 | 'Corporate Tax Avoidance: Is Tax Transparency The Solution?': A Practitioner View | Murphy R. | Accounting and Business Research | 2019 |
| 352 | Corporate Tax Avoidance: Is Tax Transparency The Solution? | Oats L., Tuck P. | Accounting and Business Research | 2019 |
| 353 | Transfer Prices And Innovation In Public Healthcare: Costing And Clinical Choices In The Nhs | Guven-Uslu P., Seal W. | Financial Accountability and Management | 2019 |
| 354 | Tax Evasion In New Disguise? Examining Tax Havens' International Bank Deposits | Menkhoff L., Miethe J. | Journal of Public Economics | 2019 |
| 355 | Global Wealth Inequality | Zucman G. | Annual Review of Economics | 2019 |
| 356 | Local Social Environment, Firm Tax Policy, And Firm Characteristics | Gao Z., Lu L.Y., Yu Y. | Journal of Business Ethics | 2019 |
| 357 | Turning Vice Into Virtue: Institutional Work And Professional Misconduct | Harrington B. | Human Relations | 2019 |
| 358 | Made In The U.S.A.? A Study Of Firm Responses To Domestic Production Incentives | Lester R. | Journal of Accounting Research | 2019 |
| 359 | Escape From The Usa: Government Debt-To-Gdp Ratio, Country Tax Competitiveness, And Us-Oecd Cross-Border M&As | Gan Y., Qiu B. | Journal of International Business Studies | 2019 |
| 360 | The Impact Of Corporate Governance On Corporate Tax Avoidance–A Literature Review | Kovermann J., Velte P. | Journal of International Accounting, Auditing and Taxation | 2019 |
| 361 | Progressive Wealth Taxation | Saez E., Zucman G. | Brookings Papers on Economic Activity | 2019 |
| 362 | Progressive Wealth Taxation | Saez E., Zucman G. | Brookings Papers on Economic Activity | 2019 |
| 363 | The Impact Of Tax And Transfer Pricing On A Multinational Firm'S Strategic Decision Of Selling To A Rival | Hsu V.N., Xiao W., Xu J. | Production and Operations Management | 2019 |
| 364 | The Effect Of International Income Shifting On The Link Between Real Investment And Corporate Taxation | Marques M., Montenegro T., Pinho C. | Journal of International Accounting, Auditing and Taxation | 2019 |

(continued)

**Table A.1** (continued)

| N | Title | Authors | Journal | Year |
|---|---|---|---|---|
| 365 | Understanding The Rise In Corporate Cash: Precautionary Savings Or Foreign Taxes | Faulkender M.W., Hankins K.W., Petersen M.A. | Review of Financial Studies | 2019 |
| 366 | How Quickly Do Firms Adjust To Optimal Levels Of Tax Avoidance? | Kim J., McGuire S.T., Savoy S., Wilson R. | Contemporary Accounting Research | 2019 |
| 367 | Trade-Offs Between Tax And Financial Reporting Benefits: Evidence From Purchase Price Allocations In Taxable Acquisitions | Lynch D., Robinson J.R., Romney M., Stomberg B., Wangerin D. | Contemporary Accounting Research | 2019 |
| 368 | Overconfidence And Corporate Tax Policy | Chyz J.A., Gaertner F.B., Kausar A., Watson L. | Review of Accounting Studies | 2019 |
| 369 | The Value Implications Of Tax Avoidance Across Countries | Tang T.Y.H. | Journal of Accounting, Auditing and Finance | 2019 |
| 370 | The Effect Of Cross-Border Group Taxation On Ownership Chains | Rünger S. | European Accounting Review | 2019 |
| 371 | Tax Planning Of Married Couples And Intra-Household Income Inequality | Buettner T., Erbe K., Grimm V. | Journal of Public Economics | 2019 |
| 372 | The Rise Of The Digital Economy: Rethinking The Taxation Of Multinational Enterprises | Gray S.J., Ting A. | Journal of International Business Studies | 2019 |
| 373 | Shifting Mne Taxation From National To Global Profits: A Radical Reform Long Overdue | McGaughey S.L., Raimondos P. | Journal of International Business Studies | 2019 |
| 374 | Taxing The Multinational Enterprise: On The Forced Redesign Of Global Value Chains And Other Inefficiencies | Foss N.J., Mudambi R., Murtinu S. | Journal of International Business Studies | 2019 |
| 375 | Does Tax Planning Affect Analysts' Forecast Accuracy? | Francis J.R., Neuman S.S., Newton N.J. | Contemporary Accounting Research | 2019 |
| 376 | The Impact Of Corporate Tax Avoidance On Board Of Directors And Ceo Reputation | Lanis R., Liu C., McClure R., Richardson G. | Journal of Business Ethics | 2019 |
| 377 | Tax Avoidance Using Hybrid Financial Instruments Among European Countries | Holtmann S. | British Tax Review | 2020 |
| 378 | Withholding Tax Planning: Should It Be Disclosed, And How Might It Be Challenged? | Gabbai S. | British Tax Review | 2020 |
| 379 | Intangibles And The Transfer Pricing Reconstruction Rules: A Case Study Of Amazon | Ting A. | British Tax Review | 2020 |

(continued)

**Table A.1** (continued)

| N | Title | Authors | Journal | Year |
|---|---|---|---|---|
| 380 | Holding Proteus: Emerging Treaty Practice On Hybrid And Fiscally Transparent Entities | Brabazon M. | British Tax Review | 2020 |
| 381 | Sasol Oil V Csars: Judicial Reinterpretation Of Legal Doctrine To Address Tax Avoidance In South Africa | Hattingh J., Titus A. | British Tax Review | 2020 |
| 382 | The Effect Of Chief Financial Officers' Accounting Expertise On Corporate Tax Avoidance: The Role Of Compensation Design | Chang C., Chen M., Lee M. | Review of Quantitative Finance and Accounting | 2020 |
| 383 | Estimating Bargaining-Related Tax Advantages Of Multinational Firms | Egger P.H., Strecker N.M., Zoller-Rydzek B. | Journal of International Economics | 2020 |
| 384 | Corporate Tax Avoidance And Debt Costs | Shevlin T., Urcan O., Vasvari F.P. | Journal of the American Taxation Association | 2020 |
| 385 | Income Shifting As Income Creation? | Selin H., Simula L. | Journal of Public Economics | 2020 |
| 386 | The Impact Of Corporate Tax Avoidance On Analyst Coverage And Forecasts | He G., Ren H.M., Taffler R. | Review of Quantitative Finance and Accounting | 2020 |
| 387 | Advertising And Tax Avoidance | Mansi S., Qi J., Shi H. | Review of Quantitative Finance and Accounting | 2020 |
| 388 | Trading Out Of Sight: An Analysis Of Cross-Trading In Mutual Fund Families | Eisele A., Nefedova T., Parise G., Peijnenburg K. | Journal of Financial Economics | 2020 |
| 389 | At A Cost: The Real Effects Of Transfer Pricing Regulations | de Mooij R., Liu L. | IMF Economic Review | 2020 |
| 390 | Tax Avoidance, Corporate Governance, And Corporate Social Responsibility: The Case Of The Egyptian Capital Market | Abdelfattah T., Aboud A. | Journal of International Accounting, Auditing and Taxation | 2020 |
| 391 | Transparency And Tax Evasion: Evidence From The Foreign Account Tax Compliance Act (Fatca) | De Simone L., Lester R., Markle K. | Journal of Accounting Research | 2020 |
| 392 | Taxation And The Allocation Of Risk Inside The Multinational Firm | Becker J., Johannesen N., Riedel N. | Journal of Public Economics | 2020 |
| 393 | Corporate Inversions: Going Beyond Tax Incentives | Col B., Liao R., Zeume S. | Review of Corporate Finance Studies | 2020 |
| 394 | Speaking Truth To Power: Twitter Reactions To The Panama Papers | Everett J., Neu D., Saxton G., Shiraz A.R. | Journal of Business Ethics | 2020 |

(continued)

**Table A.1** (continued)

| N | Title | Authors | Journal | Year |
|---|---|---|---|---|
| 395 | Tax Fairness: Conceptual Foundations And Empirical Measurement | Farrar J., Massey D.W., Osecki E., Thorne L. | Journal of Business Ethics | 2020 |
| 396 | The Moderating Role Of Internal Control In Tax Avoidance: Evidence From A Coso-Based Internal Control Index In China | Chen H., Yang D., Zhang X., Zhou N. | Journal of the American Taxation Association | 2020 |
| 397 | Allocation Of Property Rights And Technological Innovation Within Firms | Magelssen C. | Strategic Management Journal | 2020 |
| 398 | Tax-Motivated Transfer Mispricing In South Africa: Direct Evidence Using Transaction Data | Wier L. | Journal of Public Economics | 2020 |
| 399 | Unilateral Tax Reform: Border Adjusted Taxes, Cash Flow Taxes, And Transfer Pricing | Bond E.W., Gresik T.A. | Journal of Public Economics | 2020 |
| 400 | Does Private Country-By-Country Reporting Deter Tax Avoidance And Income Shifting? Evidence From Beps Action Item 13 | Joshi P. | Journal of Accounting Research | 2020 |
| 401 | Technology And Tax Systems | Agrawal D.R., Wildasin D.E. | Journal of Public Economics | 2020 |
| 402 | Deathbed Tax Planning | Erixson O., Escobar S. | Journal of Public Economics | 2020 |
| 403 | Profit Shifting And Tax-Rate Uncertainty | Delis M.D., Hasan I., Karavitis P.I. | Journal of Business Finance and Accounting | 2020 |
| 404 | Is Corporate Tax Aggressiveness A Reputation Threat? Corporate Accountability, Corporate Social Responsibility, And Corporate Tax Behavior | Baudot L., Johnson J.A., Roberts A., Roberts R.W. | Journal of Business Ethics | 2020 |
| 405 | Firm Valuation And The Uncertainty Of Future Tax Avoidance | Jacob M., Schütt H.H. | European Accounting Review | 2020 |
| 406 | Executives' Personal Tax Behavior And Corporate Tax Avoidance Consistency | Hjelström T., Kallunki J.P., Nilsson H., Tylaite M. | European Accounting Review | 2020 |
| 407 | Are Changes Of Organizational Form Costly? Income Shifting And Business Entry Responses To Taxes | Tazhitdinova A. | Journal of Public Economics | 2020 |
| 408 | Legal Environment And Corporate Tax Avoidance: Evidence From State Tax Codes | Ma M., Thomas W.B. | Journal of the American Taxation Association | 2020 |

(continued)

**Table A.1** (continued)

| N | Title | Authors | Journal | Year |
|---|---|---|---|---|
| 409 | A Methodology To Measure The Quality Of Tax Avoidance Case Studies: Findings From The Netherlands | Gunn A.F., Koch D.J., Weyzig F. | Journal of International Accounting, Auditing and Taxation | 2020 |
| 410 | The Cost Of Independence: Evidence From Companies' Decisions To Dismiss Audit Firms As Tax-Service Providers | Cook K.A., Kim K., Omer T.C. | Accounting Horizons | 2020 |
| 411 | Strategic Subsidiary Disclosure | Dyreng S.D., Hoopes J.L., Langetieg P., Wilde J.H. | Journal of Accounting Research | 2020 |
| 412 | Ceo Severance Pay And Corporate Tax Planning | Campbell J.L., Guan J.X., Li O.Z., Zheng Z. | Journal of the American Taxation Association | 2020 |
| 413 | Corporate Tax Reforms And Tax-Motivated Profit Shifting: Evidence From The Eu | Alexander A., De Vito A., Jacob M. | Accounting and Business Research | 2020 |
| 414 | Ethics Of Tax Interpretation | Ostas D.T. | Journal of Business Ethics | 2020 |
| 415 | Can Corporate Income Tax Cuts Stimulate Innovation? | Atanassov J., Liu X. | Journal of Financial and Quantitative Analysis | 2020 |
| 416 | Tax Evasion, Capital Gains Taxes, And The Housing Market | Agarwal S., Li K., Qin Y., Wu J., Yan J. | Journal of Public Economics | 2020 |
| 417 | Taxing Hidden Wealth: The Consequences Of Us Enforcement Initiatives On Evasive Foreign Accounts | Johannesen N., Langetieg P., Reck D., Risch M., Slemrod J. | American Economic Journal: Economic Policy | 2020 |
| 418 | Aggressive Tax Avoidance By Managers Of Multinational Companies As A Violation Of Their Moral Duty To Obey The Law: A Kantian Rationale | Lenz H. | Journal of Business Ethics | 2020 |
| 419 | Corporate Social Responsibility Of U.S.-Listed Firms Headquartered In Tax Havens | Lee D. | Strategic Management Journal | 2020 |
| 420 | Foreign Exchange Risk, Hedging, And Tax-Motivated Outbound Income Shifting | Deng Z. | Journal of Accounting Research | 2020 |
| 421 | Political Status And Tax Haven Investment Of Emerging Market Firms: Evidence From China | Deng Z., Sun P., Yan J. | Journal of Business Ethics | 2020 |
| 422 | Assessing Tax Risk: Practitioner Perspectives | Neuman S.S., Omer T.C., Schmidt A.P. | Contemporary Accounting Research | 2020 |

(continued)

**Table A.1** (continued)

| N | Title | Authors | Journal | Year |
|---|---|---|---|---|
| 423 | The Alignment Between Reported Profits And Real Activity In Times Of The Beps Action Plan | Schulte Sasse K., Watrin C., Weizig F. | Journal of International Accounting, Auditing and Taxation | 2020 |
| 424 | The Economic Effects Of Special Purpose Entities On Corporate Tax Avoidance | Demeré P., Donohoe M.P., Lisowsky P. | Contemporary Accounting Research | 2020 |
| 425 | How Reliably Do Empirical Tests Identify Tax Avoidance? | De Simone L., Nickerson J., Seidman J., Stomberg B. | Contemporary Accounting Research | 2020 |
| 426 | How Aggressive Tax Planning Facilitates The Diversion Of Corporate Resources: Evidence From Path Analysis | Bauer A.M., Fang J., Pittman J., Zhang Y., Zhao Y. | Contemporary Accounting Research | 2020 |
| 427 | Corporate Tax Avoidance And Mutual Fund Ownership | Doellman T., Huseynov F., Nasser T., Sardarli S. | Accounting and Business Research | 2020 |
| 428 | Tax Avoidance And Asymmetric Cost Behavior | Xu S., Zheng K. | Journal of Accounting, Auditing and Finance | 2020 |
| 429 | The Effect Of A Worldwide Tax System On Tax Management Of Foreign Subsidiaries | Kohlhase S., Pierk J. | Journal of International Business Studies | 2020 |
| 430 | Cross-Border Tax Evasion After The Common Reporting Standard: Game Over? | Casi E., Spengel C., Stage B.M.B. | Journal of Public Economics | 2020 |
| 431 | International Transfer Pricing And Tax Avoidance: Evidence From Linked Trade-Tax Statistics In The United Kingdom | Guo D., Liu L., Schmidt-Eisenlohr T. | Review of Economics and Statistics | 2020 |
| 432 | Are Less Developed Countries More Exposed To Multinational Tax Avoidance? Method And Evidence From Micro-Data | Johannesen N., Tørsløv T., Wier L. | World Bank Economic Review | 2020 |
| 433 | Information System And Corporate Income Tax Enforcement: Evidence From China | Shao Y., Xiao C. | Journal of Accounting and Public Policy | 2020 |
| 434 | Transfer Pricing Regulation And Tax Competition | Choi J.P., Furusawa T., Ishikawa J. | Journal of International Economics | 2020 |
| 435 | The Effects Of The Tax Cuts And Jobs Act Of 2017 On Defined Benefit Pension Contributions* | Gaertner F.B., Lynch D.P., Vernon M.E. | Contemporary Accounting Research | 2020 |

(continued)

**Table A.1** (continued)

| N | Title | Authors | Journal | Year |
|---|---|---|---|---|
| 436 | Reducing Tax Compliance Costs Through Corporate Tax Base Harmonization In The European Union | Barrios S., d'Andria D., Gesualdo M. | Journal of International Accounting, Auditing and Taxation | 2020 |
| 437 | An Empirical Examination Of The Influence Of E-Commerce On Tax Avoidance In Europe | Argilés-Bosch J.M., García-Blandón J., Ravenda D., Somoza A. | Journal of International Accounting, Auditing and Taxation | 2020 |
| 438 | Does Public Country-By-Country Reporting Deter Tax Avoidance And Income Shifting? Evidence From The European Banking Industry* | Joshi P., Outslay E., Persson A. | Contemporary Accounting Research | 2020 |
| 439 | Tax Avoidance And Financial Statement Readability | Nguyen J.H. | European Accounting Review | 2021 |
| 440 | Dynamics Of The Market For Corporate Tax-Avoidance Advice* | Konrad K.A. | Scandinavian Journal of Economics | 2021 |
| 441 | The Concave Association Between Tax Reserves And Equity Value | Moore R.D. | Journal of the American Taxation Association | 2021 |
| 442 | The Effect Of Tax Authority Enforcement On Earnings Informativeness | Zhao L. | European Accounting Review | 2021 |
| 443 | The Conflict Of Interest In Tax Scholarship | Raitasuo S. | Critical Perspectives on Accounting | 2021 |
| 444 | Customer And Tax Behaviour: How Customer Concentration Affect Suppliers' Tax Avoidance | Mao N., Wang J. | Accounting Forum | 2021 |
| 445 | Aggressive Tax Avoiders: U.S. Multinationals Shifting Domestic Earnings To Zero | Chen N.X., Lehmer T. | Accounting Review | 2021 |
| 446 | Tax Advisors And Tax Aggressiveness: A Bargaining Model | Elitzur R., Yaari V. | Journal of Accounting, Auditing and Finance | 2021 |
| 447 | Is Corporate Social Responsibility Related To Corporate Tax Avoidance? Evidence From A Natural Experiment | Mayberry M.A., Watson L. | Journal of the American Taxation Association | 2021 |
| 448 | Trouble In Paradise? Disclosure After Tax Haven Leaks | Schmal F., Schulte Sasse K., Watrin C. | Journal of Accounting, Auditing and Finance | 2021 |
| 449 | Identifying Different Types Of Tax Avoidance: Implications For Empirical Research | Guenther D.A., Krull L.K., Williams B.M. | Journal of the American Taxation Association | 2021 |
| 450 | Estimating And Imputing Missing Tax Loss Carryforward Data To Reduce Measurement Error | Max M.M., Wielhouwer J.L., Wiersma E. | European Accounting Review | 2021 |

(continued)

Table A.1 (continued)

| N | Title | Authors | Journal | Year |
|---|---|---|---|---|
| 451 | The Decreasing Trend In U.S. Cash Effective Tax Rates: The Role Of Growth In Pre-Tax Income | Edwards A., Kubata A., Shevlin T. | Accounting Review | 2021 |
| 452 | David Versus Goliath: The Relation Between Auditor Size And Audit Quality For U.K. Private Firms | Chen J.Z., Elemes A., Lobo G.J. | European Accounting Review | 2021 |
| 453 | The Effect Of Income Shifting On The Implied Cost Of Equity Capital: Evidence From Us Multinational Corporations | Hasan M.M., Obaydin I., Richardson G., Taylor G. | Accounting and Business Research | 2021 |
| 454 | Proprietary Costs And The Reporting Of Segment-Level Tax Expense | Deng Z., Gaertner F.B., Lynch D.P., Steele L.B. | Journal of the American Taxation Association | 2021 |
| 455 | Returnee Directors And Corporate Tax Avoidance | Chen D., Chen J., Liu L., Wang Z. | Journal of Accounting, Auditing and Finance | 2021 |
| 456 | The Effects Of Tax Avoidance News On Employee Perceptions Of Managers And Firms: Evidence From Glassdoor.Com Ratings | Lee Y., Ng S., Shevlin T., Venkat A. | Accounting Review | 2021 |
| 457 | The Tax Complexity Index–A Survey-Based Country Measure Of Tax Code And Framework Complexity | Hoppe T., Schanz D., Sturm S., Sureth-Sloane C. | European Accounting Review | 2021 |
| 458 | The Political Dynamics Of Corporate Tax Avoidance: The Chinese Experience | Chen H., Tang S., Wu D., Yang D. | Accounting Review | 2021 |
| 459 | Corporate Governance And Profit Shifting: The Role Of The Audit Committee | Delis F., Delis M.D., Karavitis P.I., Klassen K.J. | European Accounting Review | 2021 |
| 460 | Corporate Value Added Tax Avoidance | Huo Z., Rui O.M., Tang X., Zeng Y., Zhang J. | Accounting Forum | 2021 |
| 461 | Is Managerial Rent Extraction Associated With Tax Aggressiveness? Evidence From Informed Insider Trading | Gao X., Jia Y. | Review of Quantitative Finance and Accounting | 2021 |
| 462 | Capacity Rights And Full-Cost Transfer Pricing | Dutta S., Reichelstein S. | Management Science | 2021 |
| 463 | The Trouble With Instruments: The Need For Pretreatment Balance In Shock-Based Instrumental Variable Designs | Atanasov V., Black B. | Management Science | 2021 |
| 464 | International Taxation And Production Outsourcing | Lai G., Liu Y., Xiao W. | Production and Operations Management | 2021 |

(continued)

**Table A.1** (continued)

| N | Title | Authors | Journal | Year |
|---|---|---|---|---|
| 465 | Intertemporal Labor Supply Substitution? Evidence From The Swiss Income Tax Holidays | Martínez I.Z., Saez E., Siegenthaler M. | American Economic Review | 2021 |
| 466 | Fin 48 And The Tax Aggressive Behaviors Of Transnational Corporations: A Decade Later | Borkowski S.C., Gaffney M.A. | Journal of International Accounting, Auditing and Taxation | 2021 |
| 467 | Financial Constraints And Future Tax Outcome Volatility | Akamah H.T., Omer T.C., Shu S.Q. | Journal of Business Finance and Accounting | 2021 |
| 468 | Innovation And Corporate Tax Planning: The Distinct Effects Of Patents And R&D* | Cheng C.S.A., Guo P., Weng C.H., Wu Q. | Contemporary Accounting Research | 2021 |
| 469 | Corporate In-House Tax Departments* | Chen X., Cheng Q., Chow T., Liu Y. | Contemporary Accounting Research | 2021 |
| 470 | Information Processing Costs And Corporate Tax Avoidance: Evidence From The Sec'S Xbrl Mandate | Chen J.Z., Hong H.A., Kim J.B., Ryou J.W. | Journal of Accounting and Public Policy | 2021 |
| 471 | The Effect Of Tax Preparers On Corporate Tax Aggressiveness: Evidence Form The Uk Context | Assidi S., Hussainey K. | International Journal of Finance and Economics | 2021 |
| 472 | Consumer Reactions To Tax Avoidance: Evidence From The United States And Germany | Hardeck I., Harden J.W., Upton D.R. | Journal of Business Ethics | 2021 |
| 473 | Dynamic Resource Allocation With Hidden Volatility | Feng F.Z., Westerfield M.M. | Journal of Financial Economics | 2021 |
| 474 | Clearing The Bar: Improving Tax Compliance For Small Firms Through Target Setting | Al-Karablieh Y., Koumanakos E., Stantcheva S. | Journal of International Economics | 2021 |
| 475 | The Effects Of Immigration On Labour Tax Avoidance: An Empirical Spatial Analysis | Argiles-Bosch J.M., García-Blandón J., Ravenda D., Valencia-Silva M.M. | Journal of Business Ethics | 2021 |
| 476 | Stakeholder Agency Relationships: Ceo Stock Options And Corporate Tax Avoidance | Martin G.P., O'Sullivan D., Wiseman R.M., Zolotoy L. | Journal of Management Studies | 2021 |
| 477 | Asset Redeployability And Corporate Tax Avoidance | Alam N., Habib A., Hasan M.M. | Abacus | 2021 |
| 478 | Inequality, Autocracy, And Sovereign Funds As Determinants Of Foreign Portfolio Equity Flows | Kemme D.M., Parikh B., Steigner T. | Journal of Financial Research | 2021 |
| 479 | Financial Reporting Incentives, Earnings Management, And Tax Avoidance In Smes | Sánchez-Ballesta J.P., Yagüe J. | Journal of Business Finance and Accounting | 2021 |

(continued)

**Table A.1** (continued)

| N | Title | Authors | Journal | Year |
|---|---|---|---|---|
| 480 | International Taxation And Productivity Effects Of M&As | Todtenhaupt M., Voget J. | Journal of International Economics | 2021 |
| 481 | The Importance Of Irs Enforcement To Stock Price Crash Risk: The Role Of Ceo Power And Incentives | Bauer A.M., Fang X., Pittman J.A. | Accounting Review | 2021 |
| 482 | What Matters For In-House Tax Planning: Tax Function Power And Status | Ege M.S., Hepfer B.F., Robinson J.R. | Accounting Review | 2021 |
| 483 | External Auditor Responses To Tax Risk | Abernathy J.L., Finley A.R., Rapley E.T., Stekelberg J. | Journal of Accounting, Auditing and Finance | 2021 |
| 484 | Spillover Effects Of Tax Avoidance On Peers' Firm Value | Bauckloh T., Hardeck I., Inger K.K., Wittenstein P., Zwergel B. | Accounting Review | 2021 |
| 485 | The Impact Of The Oakland Ssb Tax On Prices And Volume Sold: A Study Of Intended And Unintended Consequences | Léger P.T., Powell L.M. | Health Economics (United Kingdom) | 2021 |
| 486 | Redrawing The Map Of Global Capital Flows: The Role Of Cross-Border Financing And Tax Havens | Coppola A., Maggiori M., Neiman B., Schreger J. | Quarterly Journal of Economics | 2021 |
| 487 | Income Shifting And U.S. International Trade In Goods Statistics | Deng Z., Laux R.C. | Journal of Accounting and Public Policy | 2021 |
| 488 | Financial Transparency To The Rescue: Effects Of Public Country-By-Country Reporting In The European Union Banking Sector On Tax Avoidance* | Overesch M., Wolff H. | Contemporary Accounting Research | 2021 |
| 489 | The Effect Of Disclosing Audit Quality Control Deficiencies On Non-Audit Tax Services: Evidence From Deloitte'S 2007 Pcaob Part Ii Inspection Report | Ahn J., Akamah H., Shu S.Q. | Journal of Accounting and Public Policy | 2021 |
| 490 | Did Pcaob Rules On Ethics, Independence, And Tax Services Influence Financial Reporting For Income Taxes? | Aier J.K., Cao J., Carr K.M. | Journal of Accounting and Public Policy | 2021 |
| 491 | Do Financing Constraints Lead To Incremental Tax Planning? Evidence From The Pension Protection Act Of 2006* | Campbell J.L., Goldman N.C., Li B. | Contemporary Accounting Research | 2021 |

(continued)

**Table A.1** (continued)

| N | Title | Authors | Journal | Year |
|---|---|---|---|---|
| 492 | The Association Between Auditor Provided Tax Planning And Tax Compliance Services And Tax Avoidance And Tax Risk | Chyz J.A., Gal-Or R., Naiker V., Sharma D.S. | Journal of the American Taxation Association | 2021 |
| 493 | Monitoring Or Payroll Maximization? What Happens When Workers Enter The Boardroom? | Gleason C.A., Kieback S., Thomsen M., Watrin C. | Review of Accounting Studies | 2021 |
| 494 | When Dedicated Investors Are Distracted: The Effect Of Institutional Monitoring On Corporate Tax Avoidance | Li B., Liu Z., Wang R. | Journal of Accounting and Public Policy | 2021 |
| 495 | The Internet As A Tax Haven?' | Agrawal D.R. | American Economic Journal: Economic Policy | 2021 |
| 496 | The Influence Of Organizations' Tax Avoidance Practices On Consumers' Behavior: The Role Of Moral Reasoning Strategies, Political Ideology, And Brand Identification | Matute J., Palau-Saumell R., Sánchez-Torelló J.L. | Journal of Business Ethics | 2021 |
| 497 | For Whom (And For When) Is The Firm Governed? The Effect Of Changes In Corporate Fiduciary Duties On Tax Strategies And Earnings Management | Cumming D., Tingle QC B.C., Zhan F. | European Financial Management | 2021 |
| 498 | Corporate Governance And Wealth And Income Inequality | Brou D., Chatterjee A., Coakley J., Girardone C., Wood G. | Corporate Governance: An International Review | 2021 |
| 499 | E-Commerce And Labour Tax Avoidance | Argilés-Bosch J.M., Garcia-Blandón J., Ravenda D. | Critical Perspectives on Accounting | 2021 |
| 500 | Tax Haven Incorporation And The Cost Of Capital* | Lewellen C.M., Mauler L., Watson L. | Contemporary Accounting Research | 2021 |
| 501 | Internal Information Quality And State Tax Planning* | Laplante S.K., Lynch D.P., Vernon M.E. | Contemporary Accounting Research | 2021 |
| 502 | The Unintended Consequences Of Headquarters' Involvement In Decentralized Transfer Price Negotiations: Experimental Evidence | Arnold M., Elsinger F., Rankin F.W. | Management Science | 2021 |
| 503 | Country Portfolio And Taxation: Evidence From Japan | Gu J. | Journal of Business Ethics | 2022 |

(continued)

**Table A.1** (continued)

| N | Title | Authors | Journal | Year |
|---|---|---|---|---|
| 504 | Real Effects Of Corporate Taxation: A Review | Jacob M. | European Accounting Review | 2022 |
| 505 | Tax Haven Incorporation And Financial Reporting Transparency | Lewellen C.M. | Review of Accounting Studies | 2022 |
| 506 | Profit Shifting, Returns On Foreign Direct Investments And Investment Income Imbalances | Vicard V. | IMF Economic Review | 2022 |
| 507 | Correction: Profit Shifting, Returns On Foreign Direct Investments And Investment Income Imbalances (Imf Economic Review, (2022), 10.1057/S41308-022-00178-4) | Vicard V. | IMF Economic Review | 2022 |
| 508 | Hmrc V Euromoney Institutional Investor Plc: An Example Of Tax Planning As "Icing On The Cake" | Squires S. | British Tax Review | 2022 |
| 509 | Tax Planning Under The Globe Rules | Wardell-Burrus H. | British Tax Review | 2022 |
| 510 | Pillar Two From The Perspective Of The European Union | Dourado A.P. | British Tax Review | 2022 |
| 511 | Corporate Philanthropy As A Response To Greater Tax Enforcement | Zhao L. | Accounting and Business Research | 2022 |
| 512 | Transfer Pricing: Changing Views In Changing Times | Oats L., Rogers H. | Accounting Forum | 2022 |
| 513 | The Effect Of Changes In Income Shifting On Affiliate Managers' Internal Reporting Decisions' | Klassen K.J., Ruiz C.V. | Contemporary Accounting Research | 2022 |
| 514 | The Media Representation Of Luxleaks: A Window Onto The Normative Dynamics Of Tax Avoidance From A Socio-Legal Perspective | Gendron Y., Mayer M. | Critical Perspectives on Accounting | 2022 |
| 515 | Optimal Revision Rules Of Cost-Based Transfer Prices In A Multi-Period Production Setting | Brunner M., Schäfer P. | European Accounting Review | 2022 |
| 516 | On The Dynamics Between Local And International Tax Planning In Multinational Corporations | Beuselinck C., Pierk J. | Review of Accounting Studies | 2022 |
| 517 | The Dominant Role Of Large Firms In Profit Shifting | Erasmus H., Wier L. | IMF Economic Review | 2022 |
| 518 | The Role Of Creditor Protection In Lending And Tax Avoidance | De Vito A., Jacob M. | Journal of Financial and Quantitative Analysis | 2022 |
| 519 | Compliance Challenges Of The Beps Two-Pillar Solution | Noonan C., Plekhanova V. | British Tax Review | 2022 |

(continued)

**Table A.1** (continued)

| N | Title | Authors | Journal | Year |
|---|---|---|---|---|
| 520 | Implementing Problematic Tax Regulation: Hysteresis And Bureaucratic Revolutionaries Within Tax Administrations | Oats L., Ormeño-Pérez R. | British Accounting Review | 2022 |
| 521 | Ownership Structure And Political Spending Disclosure | Adegbite E., Ali H., Nguyen T.H. | Accounting Forum | 2022 |
| 522 | Restrictions On Managerial Outside Job Opportunities And Corporate Tax Policy: Evidence From A Natural Experiment | Chen T.y., Chen Z., Li Y. | Journal of Accounting and Public Policy | 2022 |
| 523 | "Uncomfortable Territory": Personal And Organisational Values In The Tax Profession | Killian S., O'Regan P., O'Regan V. | Accounting Forum | 2022 |
| 524 | In Search Of The Owner: Regulating Through Transparency | Konovalova M., Ormeño Pérez R., Tuck P. | Critical Perspectives on Accounting | 2022 |
| 525 | When Do Firms Use One Set Of Books In An International Tax Compliance Game? | Reineke R., Weiskirchner-Merten K., Wielenberg S. | Review of Accounting Studies | 2022 |
| 526 | Foreign Employment, Income Shifting, And Tax Uncertainty | Drake K.D., Goldman N.C., Murphy F. | Accounting Review | 2022 |
| 527 | Effective Tax Planning | Schwab C.M., Stomberg B., Williams B.M. | Accounting Review | 2022 |
| 528 | Aggressive Tax Planning And Labor Investments | Goldman N.C., Lewellen C.M., Traini S. | Journal of Accounting, Auditing and Finance | 2022 |
| 529 | Financial Statement Comparability And Corporate Tax Strategy | Hong H.A., Ryou J.W., Srivastava A. | European Accounting Review | 2022 |
| 530 | The Minimum Wage And Corporate Tax Avoidance | Li X., Lim C.Y., Xu Y. | Accounting and Business Research | 2022 |
| 531 | Voluntary Vs. Mandatory: The Role Of Auditing In Constraining Corporate Tax Avoidance In Small Private Firms | Dong T., Tylaite M., Wilson R. | Accounting and Business Research | 2022 |
| 532 | Estimating Case-Based Individual And Social Learning In Corporate Tax Avoidance | Huang M., Pape A., Zhao S. | Oxford Bulletin of Economics and Statistics | 2022 |
| 533 | The Gift That Keeps On Giving: Stock Returns Around Ceo Stock Gifts To Family Members | Brown J.L., Huston G.R., Wenzel B.S. | Review of Accounting Studies | 2022 |
| 534 | Foreign Competitive Pressure And Inversions By U.S. Multinational Enterprises | Kumaraswamy A., Reyes-Peña R., Upadhyay A. | Journal of International Business Studies | 2022 |

(continued)

**Table A.1** (continued)

| N | Title | Authors | Journal | Year |
|---|---|---|---|---|
| 535 | Political Costs And Corporate Tax Avoidance: Evidence From Sin Firms | Wang C., Wilson R.J., Zhang S., Zou H. | Journal of Accounting and Public Policy | 2022 |
| 536 | Short-Sale Deregulation And Corporate Tax Aggressiveness: Evidence From The Chinese Market | Cao Y., Dong Y., Guo T., Ma D. | European Journal of Finance | 2022 |
| 537 | Chief Financial Officer Co-Option And Tax Avoidance In European Listed Firms | Allini A., Campa D., Casciello R., Ginesti G. | Journal of Accounting and Public Policy | 2022 |
| 538 | Taxes And Director Independence: Evidence From Board Reforms Worldwide | Li Q., Maydew E.L., Willis R.H., Xu L. | Review of Accounting Studies | 2022 |
| 539 | Involvement In Offshore Financial Centers And Audit Fees: Evidence From U.S. Multinational Firms | Ge W., Kim J.B., Li T., Zhang S.W. | European Accounting Review | 2022 |
| 540 | How Does Mandatory Ifrs Adoption Affect Tax Planning Decision? Evidence From Tax Avoidance Distributions | Sun H., Zhang C., Zhang J., Zhang X. | Accounting Forum | 2022 |
| 541 | Does Mandatory Disclosure Of Firm'S Tax Avoidance Position Affect Corporate Investment Efficiency? | Mehmood K., Tan H., Tao X., Wang H. | Accounting and Business Research | 2022 |
| 542 | Real Effects Of Tax Audits | Belnap A., Hoopes J.L., Maydew E.L., Turk A. | Review of Accounting Studies | 2022 |
| 543 | Tax-Related Restatements And Tax Avoidance Behavior | Adams M.T., Inger K.K., Maher J.J., Meckfessel M.D. | Journal of Accounting, Auditing and Finance | 2022 |
| 544 | Are Financial Derivatives Related To Intra-Entities' Tax Aggressiveness? Uk Evidence | Boussaidi A., Chkir I., Hussainey K., Sidhom-Hamed M. | European Financial Management | 2022 |
| 545 | Does Public Scrutiny On Corporate Tax Decisions Affect Directors? Effects Of Responsible (Irresponsible) Corporate Tax Practices On Director Reputation | Hoi C.K., Ke Y., Wu Q., Zhang H. | Accounting and Business Research | 2022 |
| 546 | Director And Officer Liability And Corporate Tax Avoidance | Khan S., Park S.J., Veliotis S., Wald J.K. | Journal of Business Finance and Accounting | 2022 |
| 547 | Reassessing Moral Legitimacy In Times Of Instability | Anesa M., Gillespie N., Petani F.J., Spee A.P. | Journal of Management Studies | 2022 |

(continued)

**Table A.1** (continued)

| N | Title | Authors | Journal | Year |
|---|---|---|---|---|
| 548 | Do Natural Disasters Affect Corporate Tax Avoidance? The Case Of Drought | Adrian C., Garg M., Pham A.V., Phang S.Y., Truong C. | Journal of Business Ethics | 2022 |
| 549 | Board Social Capital And Stock Price Crash Risk | Chen S., Jebran K., Zhang R. | Review of Quantitative Finance and Accounting | 2022 |
| 550 | Tax Evasion And Tax Avoidance | Alstadsæter A., Johannesen N., Le Guern Herry S., Zucman G. | Journal of Public Economics | 2022 |
| 551 | Wealth Inequality: A Hybrid Approach Toward Multidimensional Distributional National Accounts In Europe | Waltl S.R. | Review of Income and Wealth | 2022 |
| 552 | Languages And Corporate Tax Avoidance | Na K., Yan W. | Review of Accounting Studies | 2022 |
| 553 | Terrorism And Corporate Tax Avoidance | Moser W.J., Xu H. | Abacus | 2022 |
| 554 | Assessing The Influence Of Different Interest Groups On International Tax Policy: Evidence From The Beps Project* | Elschner C., Hardeck I. | Contemporary Accounting Research | 2022 |
| 555 | How Do Firms Use Cash Tax Savings? A Cross-Country Analysis | Green D.H., Kerr J.N. | Journal of the American Taxation Association | 2022 |
| 556 | Are Corruption And Corporate Tax Avoidance In The United States Related? | Al-Hadi A., Richardson G., Taylor G. | Review of Accounting Studies | 2022 |
| 557 | Managerial Career Concerns And Corporate Tax Avoidance: Evidence From The Inevitable Disclosure Doctrine* | Li N., Shevlin T., Zhang W. | Contemporary Accounting Research | 2022 |
| 558 | What Determines Effective Tax Rates? The Relative Influence Of Tax And Other Factors* † | Schwab C.M., Stomberg B., Xia J. | Contemporary Accounting Research | 2022 |
| 559 | The Impact Of Juror Knowledge Of Deductibility And Defendants' Tax Rates On Punitive Damages Awards: Experimental Evidence | Church B.K., Dhooge K.D.N.L., Venkataraman S. | Journal of the American Taxation Association | 2022 |
| 560 | Cultural Diversity And Corporate Tax Avoidance: Evidence From Chinese Private Enterprises | Chan K.C., Lei G., Wang W., Yu J. | Journal of Business Ethics | 2022 |
| 561 | Climate Risk And Corporate Tax Avoidance: International Evidence | Chen Z., Li D., Ni Y., Yang S. | Corporate Governance: An International Review | 2022 |

(continued)

**Table A.1** (continued)

| N | Title | Authors | Journal | Year |
|---|---|---|---|---|
| 562 | The Effect Of Foreign Institutional Ownership On Corporate Tax Avoidance: International Evidence | Hasan I., Kim I., Teng H., Wu Q. | Journal of International Accounting, Auditing and Taxation | 2022 |
| 563 | The Incidence, Valuation, And Management Of Tax-Related Reputational Costs: Evidence From A Period Of Protest | Goodman T.H., Hoffman P.J., Schwab C.M. | Journal of the American Taxation Association | 2022 |
| 564 | Ceo Sports Hobby And Firms' Tax Aggressiveness | Luo S., Shevlin T., Shi L., Shih A. | Journal of the American Taxation Association | 2022 |
| 565 | The Impacts Of Inventory In Transfer Pricing And Net Income: Differences Between Traditional Accounting And Throughput Accounting | Antunes T.d.S., da Silva Stefano G., Lacerda D.P., Piran F.S., Wolf Motta Morandi M.I. | British Accounting Review | 2022 |
| 566 | Entity Structure And Taxes: An Analysis Of Embedded Pass-Through Entities | Agarwal A., Chen S., Mills L.F. | Accounting Review | 2022 |
| 567 | Family Firms, Tax Avoidance, And Socioemotional Wealth: Evidence From Tax Reform In Taiwan | Kuo C.S. | Review of Quantitative Finance and Accounting | 2022 |
| 568 | Ceo Duality And Tax Avoidance: Empirical Evidence From Greece | Kolias G., Koumanakos E. | Journal of International Accounting, Auditing and Taxation | 2022 |
| 569 | Is Tax Aggressiveness Associated With Tax Litigation Risk? Evidence From D&O Insurance | Donelson D.C., Glenn J.L., Yust C.G. | Review of Accounting Studies | 2022 |
| 570 | How Do Most Low Etr Firms Avoid Paying Taxes? | Christensen D.M., Kenchington D.G., Laux R.C. | Review of Accounting Studies | 2022 |
| 571 | Behavioral Economics Of Accounting: A Review Of Archival Research On Individual Decision Makers* | Hanlon M., Yeung K., Zuo L. | Contemporary Accounting Research | 2022 |
| 572 | Decomposing Multinational Corporations' Declining Effective Tax Rates | Garcia-Bernardo J., Janský P., Tørsløv T. | IMF Economic Review | 2022 |
| 573 | Expected Economic Growth And Investment In Corporate Tax Planning | Kim J., McGuire S., Savoy S., Wilson R. | Review of Accounting Studies | 2022 |
| 574 | The Effect Of Accounting For Income Tax Uncertainty On Tax-Deductible Loss Accruals For Private Insurers | Cheng J., Chow T., Lin T.T., Ng J. | Journal of Risk and Insurance | 2022 |

(continued)

**Table A.1** (continued)

| N | Title | Authors | Journal | Year |
|---|---|---|---|---|
| 575 | Offshore Profit Shifting And Aggregate Measurement: Balance Of Payments, Foreign Investment, Productivity, And The Labor Share | Guvenen F., Mataloni R.J., Rassier D.G., Ruhl K.J. | American Economic Review | 2022 |
| 576 | Transfer Pricing Comparables: Preferring A Close Neighbor Over A Far-Away Peer? | Beuselinck C., Gonnet S., Petutschnig M., Roques T., Steens B. | Journal of International Accounting, Auditing and Taxation | 2022 |
| 577 | Valuation Implications Of Socially Responsible Tax Avoidance: Evidence From The Electricity Industry | Inger K., Stekelberg J. | Journal of Accounting and Public Policy | 2022 |
| 578 | Trading Offshore: Evidence On Banks' Tax Avoidance* | Langenmayr D., Reiter F. | Scandinavian Journal of Economics | 2022 |
| 579 | Do Information Sources Matter In Corporate Tax Avoidance? The Roles Of Peer Effects And Director Interlocks | Liao Y.H., Sang T.S., Tsai Y.T. | Review of Quantitative Finance and Accounting | 2022 |
| 580 | The Global Minimum Tax | Johannesen N. | Journal of Public Economics | 2022 |
| 581 | Economic Integration And Agglomeration Of Multinational Production With Transfer Pricing | Kato H., Okoshi H. | International Economic Review | 2022 |
| 582 | Globalization And U.S. Corporate Tax Policies: Evidence From Import Competition | Chen T., Lin C., Shao X. | Management Science | 2022 |
| 583 | Corporate Income Taxation And Inequality: Review And Discussion Of Issues Raised In The Triumph Of Injustice–How The Rich Dodge Taxes And How To Make Them Pay (2019) | Faccio T., Iacono R. | Review of Income and Wealth | 2022 |
| 584 | Ethical Reasoning In Tax Practice: Law Or Is There More? | Doyle E., Frecknall-Hughes J., Summers B. | Journal of International Accounting, Auditing and Taxation | 2022 |
| 585 | Corporate Social Responsibility And Tax Management: The Moderating Effect Of Beliefs About Corporate Tax Duty | Davis A.B., Moore R.D., Rupert T.J. | Journal of the American Taxation Association | 2022 |
| 586 | Real Effects Of Private Country-By-Country Disclosure | De Simone L., Olbert M. | Accounting Review | 2022 |
| 587 | Real Responses To Anti-Tax Avoidance: Evidence From The Uk Worldwide Debt Cap | Bilicka K., Qi Y., Xing J. | Journal of Public Economics | 2022 |

(continued)

**Table A.1** (continued)

| N | Title | Authors | Journal | Year |
|---|---|---|---|---|
| 588 | Brand Capital And Stock Price Crash Risk | Hasan M.M., Richardson G., Taylor G. | Management Science | 2022 |
| 589 | Time Orientation In Languages And Tax Avoidance | Cheng C.S.A., Kim J., Rhee M., Zhou J. | Journal of Business Ethics | 2022 |
| 590 | Corporate Tax Behavior And Political Uncertainty: Evidence From National Elections Around The World | Li Q., Maydew E.L., Willis R.H., Xu L. | Journal of Business Finance and Accounting | 2022 |
| 591 | Tax Incidence And Tax Avoidance* | Dyreng S.D., Jacob M., Jiang X., Müller M.A. | Contemporary Accounting Research | 2022 |
| 592 | Beyond Borders: Uncertainty In Supragovernmental Tax Enforcement And Corporate Investment | Fox Z.D., Jacob M., Wilde J.H., Wilson R.J. | Accounting Review | 2022 |
| 593 | Are Tax Havens And Offshore Financial Centers Cracked Down On? A Study On The International Standard Of Exchange Of Information On Request | Li Y., Ma M. | Accounting Review | 2022 |
| 594 | Capital Round-Tripping: Determinants Of Emerging Market Firm Investments Into Offshore Financial Centers And Their Ethical Implications | Brouthers K.D., Karhunen P., Ledyaeva S. | Journal of Business Ethics | 2022 |
| 595 | The Unintended Consequence Of Land Finance: Evidence From Corporate Tax Avoidance | Chen T., Tan Y., Wang J., Zeng C. | Management Science | 2022 |
| 596 | Corporate Governance And Tax Avoidance: Evidence From U.S. Cross-Listing | Chen R., Ghoul S.E., Guedhami O., Wang H., Yang Y. | Accounting Review | 2022 |
| 597 | Political Economy Of Cross-Border Income Shifting: A Protection Racket Approach | Ananyev M. | Journal of Comparative Economics | 2022 |
| 598 | Twitter-Based Social Accountability Processes: The Roles For Financial Inscriptions-Based And Values-Based Messaging | Neu D., Saxton G.D. | Journal of Business Ethics | 2022 |
| 599 | The Nonlinear Relation Between Effective Tax Rates And Tax-Related Restatements | Cheng C., Hennes K.M., Sapkota P. | Accounting Horizons | 2022 |
| 600 | Narrative Disclosure Tone: A Review And Areas For Future Research | Abdelfattah T., Bassyouny H., Tao L. | Journal of International Accounting, Auditing and Taxation | 2022 |

(continued)

**Table A.1** (continued)

| N | Title | Authors | Journal | Year |
|---|---|---|---|---|
| 601 | The Legalistic Organizational Response To Whistleblowers' Disclosures In A Scandal: Law Without Justice? | Ouriemmi O. | Journal of Business Ethics | 2023 |
| 602 | Taxing Multinational Income Based On Value Creation Versus Value Realization: An Industry Perspective | Sansing R. | Review of Accounting Studies | 2023 |
| 603 | Home Or Away? Profit Shifting With Territorial Taxation | Langenmayr D., Liu L. | Journal of Public Economics | 2023 |
| 604 | Twitter-Based Social Accountability Callouts | Neu D., Saxton G.D. | Journal of Business Ethics | 2023 |
| 605 | Firm-Level Political Risk And Corporate Tax Avoidance | Hossain M., Lobo G.J., Mitra S. | Review of Quantitative Finance and Accounting | 2023 |
| 606 | Corporate Social Responsibility And Corporate Tax Avoidance: The Channel Effect Of Consumer Awareness | Benlemlih M., Jaballah J., Peillex J., Schochet S. | Journal of Business Finance and Accounting | 2023 |
| 607 | Strategic Reaction And Tax Avoidance: Evidence From The Effect Of Large Ipos On Peers | Chen H.(., Francis B.B., Wu Q., Zhao Y. | British Accounting Review | 2023 |
| 608 | Does Corporate Tax Planning Mitigate Financial Constraints? Evidence From China | Makosa L., Sitsha L., Sun J., Yang J., Yin F. | International Journal of Finance and Economics | 2023 |
| 609 | Managerial Political Power And The Reallocation Of Resources In The Internal Capital Market | Keum D.D. | Strategic Management Journal | 2023 |
| 610 | How Does The Market For Corporate Control Impact Tax Avoidance? Evidence From International M&A Laws | Hu J., Li S., Shevlin T. | Review of Accounting Studies | 2023 |
| 611 | Anti-Mafia Police Actions, Criminal Firms, And Peer Firm Tax Avoidance | Chircop J., Fabrizi M., Malaspina P., Parbonetti A. | Journal of Accounting Research | 2023 |

www.ingramcontent.com/pod-product-compliance
Ingram Content Group UK Ltd.
Pitfield, Milton Keynes, MK11 3LW, UK
UKHW022210051026
15624UKWH00001B/100
*9783031517648*